Development and Application of Biomedical Titanium Alloys

Edited by

Liqiang Wang

Shanghai Jiao Tong University, Shanghai, PR China

&

Lai-Chang Zhang

Edith Cowan University, Perth WA , Australia

Development and Application of Biomedical Titanium Alloys

Editor: Liqiang Wang and Lai-Chang Zhang

ISBN (Online): 978-1-68108-619-4

ISBN (Print): 978-1-68108-620-0

General:

1. Any dispute or claim arising out of or in connection with this License Agreement or the Work (including non-contractual disputes or claims) will be governed by and construed in accordance with the laws of the U.A.E. as applied in the Emirate of Dubai. Each party agrees that the courts of the Emirate of Dubai shall have exclusive jurisdiction to settle any dispute or claim arising out of or in connection with this License Agreement or the Work (including non-contractual disputes or claims).

2. Your rights under this License Agreement will automatically terminate without notice and without the need for a court order if at any point you breach any terms of this License Agreement. In no event will any delay or failure by Bentham Science Publishers in enforcing your compliance with this License Agreement constitute a waiver of any of its rights.

3. You acknowledge that you have read this License Agreement, and agree to be bound by its terms and conditions. To the extent that any other terms and conditions presented on any website of Bentham Science Publishers conflict with, or are inconsistent with, the terms and conditions set out in this License Agreement, you acknowledge that the terms and conditions set out in this License Agreement shall prevail.

Bentham Science Publishers Ltd.
Executive Suite Y - 2
PO Box 7917, Saif Zone
Sharjah, U.A.E.
Email: subscriptions@benthamscience.org

**BENTHAM
SCIENCE**

CONTENTS

PREFACE

Titanium and its alloys have been widely used as biomedical implant materials due to their low density, good mechanical properties, superior corrosion resistance and biocompatibility when compared with other metallic biomaterials such as Co–Cr alloys and stainless steels. Recently, β-type titanium alloys have been increasingly considered as excellent implant materials because of the remarkable combination of high strength-to-weight ratio, good fatigue resistance, relatively low Young's modulus, good biocompatibility and high corrosion resistance. Compared to the conventional (α+β)-type titanium biomaterials such as Ti-6Al-4V, β-type titanium alloys containing non-toxic constituents possess Young's modulus (~60 GPa) closer to that of bone (10-30 GPa). From the viewpoint of practical biomedical application, it is highly desirable to obtain β-type titanium alloys. Although biomedical titanium and its alloys have been widely used, extensive endeavors have also been made in the development of new β-type titanium alloys. In addition, surface modification of titanium alloys for biomedical applications, and manufacturing of titanium alloys by new technologies (such as by two common metal additive manufacturing technologies, *i.e.* selective laser melting and electron beam melting), *etc.* are also being done. This book covers the very recent progresses in new alloy development, surface modification and additive manufacturing of biomedical titanium. This book covers the recent progress of biomedical titanium alloys in terms of the processing, microstructure, mechanical properties and corrosion properties.

Chapter 1 overviews the application of biomedical titanium alloys and recent development process of new medical grade titanium alloys in terms of composition design, biocompatibility and shape memory effect *etc.*

Chapter 2 describes the microstructure and mechanical properties of TiNbTaZr titanium alloy. The results show that the TiNbTaZr β titanium alloy with lower elastic modulus and non-toxic alloying elements has much more important application in biomedical field. The Ti-35Nb-2Ta-3Zr β titanium alloy is studied and the alloy has the complex properties of lower elastic modulus, high strength, high elongation and excellent shape memory effect. Compared with direct rolling, cross rolling is beneficial to the isotropic of the microstructure and mechanical properties.

Chapter 3 presents the microstructure and mechanical properties of a serial of novel β-type Ti-Fe-based alloys developed very recently by using the DV-Xα molecular orbital method. The influence of Ta, Fe and Nb contents on the phase transformation, β phase stability and microstructural evolution of the alloys are investigated. The resulting mechanical properties of the alloys are evaluated and compared with those of the widely used biomaterials to ascertain their suitability for orthopaedic application. The results suggest that among the designed alloys, Ti-10Fe-10Ta, Ti-7Fe-11Nb and Ti-11Nb-9Fe presents the best combination of mechanical properties making them more desirable than the commonly used CP-Ti and Ti-6Al-4V materials for implant application. It is demonstrated that through proper alloy design it is possible to design new Ti alloys with favourable properties better than CP-Ti and Ti-6Al-4V alloys for orthopaedic application.

Chapter 4 illustrates the progress in processing, microstructure and properties of biomedical titanium alloys manufactured by selective laser melting (SLM). Although titanium alloys have exhibited combination of a serial of excellent properties, the hard machinery and/or high cost of materials removal arising from the conventional manufacturing processes hinder various potential applications of titanium alloys. Emerging additive manufacturing techniques, such as SLM, are providing the ideal platform for creating titanium components especially with complex dimensions. This chapter reviews the recent progress of the SLM-produced titanium alloys and composites as well as porous structures in terms of processing, microstructure and properties.

Chapter 5 shows the microstructure and mechanical behavior of porous titanium alloys manufactured by electron beam melting of (EBM). This chapter discusses the influence of porosity variation on the mechanical properties of the β-type Ti-24Nb-4Zr-8Sn alloy porous samples, in terms of Young's modulus, super-elastic property, strength and fatigue properties. The relationship between the misorientation angle between adjacent grains and the fatigue crack deflection behaviors are also discussed. Compared with Ti-6Al-4V samples having the same porosity level, the EBM-produced β-type Ti-24Nb-4Zr-8Sn porous components exhibit a higher normalized fatigue strength owing to super-elastic property, greater plastic zone ahead of the fatigue crack tip and the crack deflection behavior. The EBM-produced Ti-24Nb-4Zr-8Sn components exhibit more than twice the strength-to-modulus ratio of porous Ti-6Al-4V counterparts.

Chapter 6 describes the preparation, microstructure and mechanical properties of NiTi-Nb porous titanium alloy. The results show that there are 4 diffusion layers, including Nb foil, NiTi-Nb eutectic region, pre-eutectic NiTi region and NiTi matrix at the junction of NiTi wire and Nb foil. In addition, a martensite phase was found inside the NiTi matrix. In the NiTi-Nb eutectic diffusion layer, niobium is inhomogeneously distributed, forming a rod-like Nb-rich phase, facetted Ti-rich phase and stripe-like or equiaxed NiTi-Nb eutectic structure. The experimental study on the microstructure and mechanical properties of interface between NiTi and Nb will help to understand the advantages of using Nb as the NiTi alloy connection material, provide theoretical support for preparation of NiTi lattice materials, and promote NiTi alloy more extensive application.

Chapter 7 presents the surface modification of biomedical Ti-6Al-4V alloy by friction stir processing (FSP). Aiming at solving the problem of poor surface wear properties and improving biocompatibility of Ti-6Al-4V alloy, FSP is applied to fill with TiO_2 powder in the groove to realize surface modification and build nano-sized composite biomedical material. Change in microstructure and its relationship with mechanical performance such as hardness will be discussed. A series of experiments in biology, including cytotoxicity test, cell culture, adhesion, proliferation and alkaline phosphatase activity is carried out to verify bio-compatibility of the material, compared with original Ti-6Al-4V. The improved material is expected to provide a better environment for cells to grow.

Chapter 8 shows the very recent progress in the corrosion behaviour of titanium alloys manufactured by selective laser melting (SLM). The different manufacturing technology of titanium alloys has a substantial impact on its performance, like that the mechanical properties of Ti-6Al-4V manufactured by selective laser melting (SLM), exhibiting comparable and even better mechanical properties than the counterpart produced by traditional manufacturing methods. Yet, the understanding of the corrosion behavior of SLM-produced materials is unknown. This chapter reviews the recent progress of the corrosion behavior of SLM-produced Ti-6Al-4V and Ti-TiB composite. The corrosion characteristics and mechanisms are discussed in detail.

Finally, we wish to thank all the authors for their contributions to this book.

Dr. Liqiang Wang
Shanghai Jiao Tong University,
Shanghai,
PR China

&

Dr. Lai-Chang Zhang
Edith Cowan University,
Joondalup,
Australia

List of Contributors

Chengjian Zhang State Key Laboratory of Metal Matrix Composites, Shanghai Jiao Tong University, Shanghai, PR China

Junxi Zhang Shanghai Key Laboratory of Material Protection and Advanced Material in Electric Power, Shanghai University of Electric Power, Shanghai, PR China

Lai-Chang Zhang School of Engineering, Edith Cowan University, Perth, WA, Australia

Liqiang Wang State Key Laboratory of Metal Matrix Composites, Shanghai Jiao Tong University, Shanghai, PR China

Wei Huang State Key Laboratory of Metal Matrix Composites, Shanghai Jiao Tong University, Shanghai, PR China

Wei Zhang State Key Laboratory of Metal Matrix Composites, Shanghai Jiao Tong University, Shanghai, PR China

Xueting Wang State Key Laboratory of Metal Matrix Composites, Shanghai Jiao Tong University, Shanghai, PR China

Yujing Liu School of Engineering, Edith Cowan University, Perth, WA, Australia

Zihao Ding State Key Laboratory of Metal Matrix Composites, Shanghai Jiao Tong University, Shanghai, PR China

CHAPTER 1

Application of Biomedical Titanium Alloys

Liqiang Wang[1,*] and **Lai-Chang Zhang**[2]

[1] *State Key Laboratory of Metal Matrix Composites, Shanghai Jiao Tong University, No. 800 Dongchuan Road, Shanghai 200240, PR China*

[2] *School of Engineering, Edith Cowan University, 270 Joondalup Drive, Joondalup, Perth, WA, 6027, Australia*

Abstract: Titanium alloys have been widely used in medical or dental applications due to their superior biocompatibility, high strength and corrosion-resistant as well as low modulus relative to other implantable metals. In order to meet the stringent medical regulations and the advancement of bioengineering, material scientists have therefore designed a series of advanced titanium alloys. This chapter aims at reviewing the development process of new medical grade titanium alloys in terms of composition design, biocompatibility and shape memory effect *etc*.

Keywords: Biocompatibility, Shape memory effect, Titanium alloys.

INTRODUCTION

With the high demand of healthcare services in general population, the cost in healthcare expenses continuously increased globally [1]. Because of the excessive demand of medical care in major rural areas, the current supply in local never meets their needs. Therefore, the import of high tech medical devices and advanced implantable biomaterials seem to be a way-out. At the same time the local material scientists have endeavored to invent new generation of biomaterials through the advancement of material technology [2, 3].

Titanium alloys serving as implantable materials have been applied in medical and dental applications for over 70 years. Currently, the most commonly used titanium alloys are pure titanium and Ti-6Al-4V as well as Ti-6Al-7Nb alloy. In spite of the poor corrosion resistance of pure titanium in physiological environment, it is also mechanically weaker and easily wear-out. Consequently, the use of pure titanium is limited in load bearing implants except in dental cosmetology [4].

---

* **Corresponding author Liqiang Wang:** State Key Laboratory of Metal Matrix Composites, School of Materials Science and Engineering, Shanghai Jiao Tong University, No. 800 Dongchuan Road, Shanghai 200240, P.R. China; Tel: 8602134202641; Fax: 8602134202749; E-mail: wang_liqiang@sjtu.edu.cn

Not only does the Ti-6Al-4V alloy contain toxic elements of Al and V, but also the Young's modulus of this alloy is much higher than that of human bones. These undesired properties might complicate the clinical outcomes due to stress shielding effect, bone resorption, and even implant loosening [5]. Hence, the recent trend of β titanium alloy research has focused on biological and mechanical modifications such as the improvement of bio functionalities for superior tissue-integration and the fabrication of, low modulus Ti alloys for avoiding stress shielding [6, 7].

α+β TITANIUM ALLOYS IN BIOMEDICAL APPLICATIONS

Few metals with excellent mechanical properties and corrosion resistance have been widely used in hard tissue replacement such as total hip and total knee arthroplasties, artificial intervertebral disc, pedicle screw and other instrumented spinal arthrodesis in spinal surgeries as well as dental implants. Other popular applications include intravascular stents, catheters, orthodontics arch wires and cochlear implants *etc*. The details are summarized in Table **1**.

Table 1. Traditional metallic materials for biomedical application.

Alloys	Phase	Countries	Integrated Properties	Applications/ Standards
Pure Titanium TA1ELI, TA1, TA2, TA3, TA4	α	Various	$R_m \geq 200 \sim 580MPa$, $A_s \geq 15 \sim 30\%$, $Z \geq 25 \sim 30\%$, $E = 100GPa$	Orthopedics, dentistry/ International standard, Chinese national standard (Widely used)
Ti-6Al-4V	α+β	Various	$R_m \geq 825 \sim 930MPa$, $A_s \geq 15 \sim 25\%$, $Z \geq 25\%$, $E = 120GPa$	Orthopedics, dentistry/ International standard, Chinese national standard (Widely used)
Ti-6Al-7Nb	α+β	Swiss-invented Various	$R_m \geq 750MPa$, $A_s \geq 12\%$, $E = 106GPa$	Orthopedics, dentistry/ International standard, Chinese national standard (Widely used)
Ti-5Al-2.5Fe	α+β	German-invented Various	$R_m \geq 1020MPa$, $A_s \geq 10\%$, $E = 112GPa$	Orthopedics/ International standard, Chinese national standard (Discarded)
Ti-2Al-2Mo-2Zr	near α	China invented	$R_m \geq 750MPa$, $A_s \geq 25\%$, $E = 105GPa$	Orthopedics, dentistry, limited use in ocean engineering

(Table 1) contd.....

Alloys	Phase	Countries	Integrated Properties	Applications/ Standards
TiNi	Martensite	Various	$R_m \geq 1200 MPa$, $A_s \geq 25\%$, $E \approx 40 GPa$	Orthopedics, dentistry, vascular interventions/ International standard, Chinese national standard (Widely used)

Most of the elements *e.g.* Fe, Cr, Co, Ni, Ti, Ta, Nb, Mo, W in the current implantable metals served as trace elements in human body. Indeed, small amount of these trace elements are crucial to human metabolism. For example, Fe element can keep the normal function of red blood cells, and Co element can maintain vitamin B12 synthesis [8]. In general, metallic implants might decay under *in vivo* condition that triggers the release of the trace elements in addition to the weakening effect of implants. These undesired effects might jeopardize the biological functions of surrounding tissues as well as organs when occurred at certain time [9 - 11]. Therefore, this is the reason why the improvement of metal biocompatibility is of the most important work in the field. The early research of titanium alloys for biomedical applications could date back to the early 1940s. Bothe *et al.* found that pure titanium had no adverse reaction with the bony tissue of rat, and which was the first pure metal introduced to this field [12]. Leventhal *et al.* then proved the excellent biocompatibility of pure titanium ten years after Bothe's study [13]. In the early 1960-ties, Branemark investigated the blood microcirculation in rabbit tibiae with a titanium chamber, and noticed that metal and bone were perfectly integrated without rejection [14]. Since then, pure titanium has become a popular implantable material. Though pure titanium has superior corrosion resistance under physiological environment. This metal is not recommended for load bearing implant due to its low mechanical strength and poor wear resistance. Therefore, it has been limited to dental restoration and non-weight bearing implants in bone surgeries. Alternatively, Ti-6Al-4V alloy has been widely adopted in hip and knee joint surgeries because of its high mechanical properties and better processability [4]. Furthermore, Ti-3Al-2.5V alloy is also considered as the bone substitute for thigh bone and shine bone replacements [15]. In addition to the poor corrosion wear resistance, these Ti based alloys consist of Al and V elements in which V is considered a highly toxic element than Ni and Cr to human body [16]. In the literatures, tiny amount of V in bone, liver, kidney and spleen would significantly interfere the phosphates metabolism of cells, thereby directly affecting the Na+, K+, Ca+, and H+ and reaction with ATP enzyme. Without considering the toxicity of V element, the literatures reported that another element of Al in these alloys not only damaged the organs, but might also potentially result to osteomalacia, anemia, and neurological disorders due to the accumulation of aluminum salts *in vivo*.

In the middle 1980s, Switzerland and Germany had jointly developed the second generation of V-free $\alpha+\beta$ titanium alloys *i.e.* Ti-6Al-7Nb and Ti-5Al-2.5Fe alloys [17, 18]. These new materials were accepted by international standard of biomedical materials and then quickly translated to clinical practice. However, these alloys still consist of toxic elements of Al and Fe. Additionally, the Young's modulus of these materials is still far away from that of human cortical and cancellous bones. Therefore, post-operative stress shielding effect likely happens, thereby resulting to implants failure due to implant loosening and cracking [19 - 23]. To overcome these complications, materials scientists have endeavored to develop highly biocompatible, lower modulus and mechanically match β titanium materials.

TiNi alloys have a special mechanical property named shape memory effect (SME) that enables the recovery from deformation after being heated. The shape memory effect was firstly discovered by Buehler and Wiley from the US Naval Ordnance Laboratory in 1963 [24]. A one-to-one atomic ratio (equiatomic) in NiTi alloy (Nitinol) provides excellent shape memory effect at particular temperature. If the plastic deformation occurs below the transformation, the metal can recover by itself when heats up beyond the transformation temperature. In general, the shape memory effect is relative to the diffusion of less martensitic phase transformation, namely thermoelastic martensitic phase transformation caused by ordering transformation from parent phase to martensitic phase [25]. Another unique characteristic of NiTi alloy is super-elasticity (SE). As shown in Fig. (1), when strain increases beyond to the initial elastic zone, the stress will be maintained at the same level despite of the strain increased. While unloading the NiTi material, the stress will be then maintained at lower level of stress as the strain reduced. With further reduction of strain, the NiTi wire recovers to initial length without any permanent deformation. Indeed, the super-elastic property of NiTi has been widely applied in dental arch wire. The studies demonstrated that the clinical outcome was much better than the correction with the use of traditional stainless steel arch wire.

β TITANIUM ALLOYS FOR BIOMEDICAL APPLICATIONS

Current Development of β Titanium Alloys

Recently, the invention of new β titanium alloys with low modulus has attracted a lot of attentions in particular to the countries such as USA, Japan, South Korea and China. The most popular β titanium alloys include Ti-Mo, Ti-Nb, Ti-Ta or Ti-Zr that have been widely applied in biomedical fields [26 - 28]. Among the other commonly used titanium alloys, these newly fabricated β titanium alloys have lower modulus and higher strength due to the contribution of the aforementioned

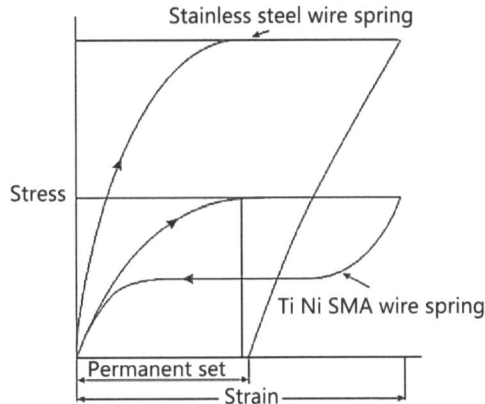

Fig. (1). Schematic illustration of the stainless steel wire and TiNi SMA wire springs for orthodontic archwire behavior.

elements. Particularly, the Ti-Nb matrix alloy with the combination of nontoxic elements including Nb, Ta and Zr enables not only the lowest modulus among the others, but also excellent shape memory effect. Therefore, this new alloy has great potential in clinical applications and is a highlighted biomedical metal in the field today. Some of the Ti-Nb matrix metastable alloys have already been used practically for example Ti-13Nb-13Zr [29], Ti-35Nb-5Ta-7Zr [30], Ti-29Nb-13Ta-4.6Zr [31], and Ti-34Nb-9Zr-8Ta [32]. Kim *et al.* had investigated the martensitic transformation, shape memory effect and super-elasticity of Ti-Nb binary system [33]. The results suggested that Ti-(15%-35%)Nb alloys performed shape memory effect and super-elasticity in which these two unique properties were attributed by α'' martensitic transformation. Moreover, the phase transformation strain and temperature decreased linearly when the Nb concentration increased. When Nb concentration contributed to 20% to 28%, the Ms decreased 40k with 1% of Nb in atomic weight increased. Also, Niinomi *et al.* invented another β titanium - Ti-29Nb-13Ta-4.6Zr (TNTZ) alloy [34]. The wire made of this new alloy displayed the mechanical strength around 700MPa to 800MPa and the elongation around 5%. Furthermore, the Young's modulus presented around 50GPa to 55GPa and the largest elastic strain was around 1.4%. Chinese Northwest Institute of Nonferrous Metals designed and developed a "relatively low cost"β titanium alloy TLM (Ti-Zr-Sn-Mo-Nb) [35, 36]. Hao *et al.* also investigated new β titanium - Ti2448 (Ti-24Nb-4Zr-7.9Sn) alloy, which the tensile strength was about 900MPa and its average Young's modulus was even less than 20GPa [37]. Wang *et al.* and his associates had also built new β (Ti-35Nb-2Ta-3Zr) alloys, of which the modulus ranged from 40GPa to 50GPa and with the strength 900MPa. In general, they investigated the microstructure and deformation characteristics of cold rolled Ti-35Nb-2Ta-3Zr alloys. Unexpectedly, it was found that the stress-induced α'' martensite and deformation of twins occurred, when the deformation ratio reached to 20%. The bulk deformation of twins was the main mechanism of plastic deformation. However, when the deformation ratio reached to 40%, dislocation glide became the main deformation mechanism. While the deformation ratio went to 99%, the better strength, larger elongation and lower modulus could be relatively easily obtained. At the same time, the super-elasticity of cold rolled TiNbTaZr alloy was found. When the works of cross rolled TiNbTaZr alloy were completed, nano crystals were then obtained under 99% deformation ratio [38 - 41].

Design of β Titanium Alloys

To develop high performance biomedical titanium alloys, pre-designing and relative mechanical properties calculation are of the most important points to consider. First of all, researchers may count on d-electron alloy design method to design their desired materials by calculating the orbital parameters B_o and M_d of Ti and addition elements. One is the bond order (hereafter referred to as B_o) which is a measure of the covalent bond strength between Ti and an alloying element. The other is the metal d-orbital energy level (M_d) which correlates with the electronegativity and the metallic radius of elements. In general, phases with lower M_d value are more stable and higher B_o value is more suitable for using solid solution strengthening method. While applying to plastic deformation method, slipping deformation is preliminary in thermo-stabilized β titanium alloy. Slipping, twin, and martensitic deformation might all appear in metastable β phase. As a result, the characteristic of β titanium alloy can be adjusted in a relatively wide range by controlling the microstructure of the material.

Furthermore, the value of M_d should be controlled between 2.35 to 2.45. And the value of B_o can be decided between 2.75 to 2.85 while designing metastable β titanium alloy. Consequently, the content of alloying elements should be strictly controlled in particular to those elements, which make β phase more stable.

Biocompatibility of β Titanium Alloys

In addition to excellent biomechanical properties, the β titanium alloys have also presented brilliant biocompatibility. Eisenbart *et al.* had initiated a study to estimate the biocompatibility of the alloying elements in their β titanium alloy and pure metal (Class 2) and AISI 316L stainless steel serving as control groups [42]. In the *in vitro* experiments, MC3T3-E1 cells and FM7373 cells were directly cultured on the samples *e.g.* pure metal such as Al, Nb, Mo, Ta, Zr, and 316L stainless steel. Fig. (**2**) suggested the proliferation, mitochondrial activity, and cell volume after 7 days of culture. The results indicated that the cells were tolerated well with the element of Nb rather than Ta, Ti, Zr and Al.

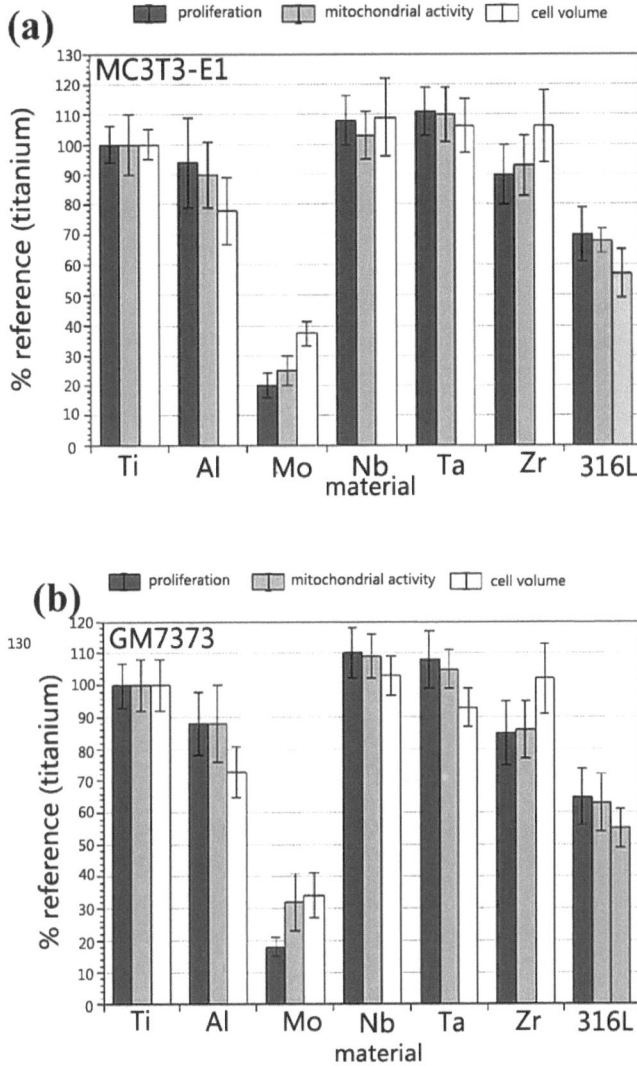

Fig. (2). Proliferation, mitochondrial activity and volume of: (**a**) MC3T3-E1 cells and (**b**) GM7373 cells after 7 days in direct contact with cp-Ti, Al, Nb, Mo, Ta, Zr and 316 L slices [42].

Niinomi and his group also studied the cytotoxicity of Ti-29Nb-13Ta-4.6Zr (TNTZ), Ti-29Nb-13Zr-2Cr, Ti-29Nb-15Zr-1.5Fe, Ti-29Nb-10Zr-0.5Si, Ti-29Nb 10Zr-0.5Cr-0.5Fe, and Ti-29Nb-18Zr-2Cr-0.5Si by MTT assay (all developed from TNTZ) [43, 44]. The overall results suggested that all those β titanium alloys exhibited good cytocompatibility. The cell viability of Ti-29Nb-13Ta-4.6Zr alloy was close to that of pure titanium, but was inferior than that of Ti-6Al-4V alloy. With respect to the *in vivo* animal test, Zhang *et al.* carried out the biocompatibility test for two kinds of β titanium alloys, namely Ti-(3-6)Zr-

(2-4)Mo-(24-27)Nb (TLE), and Ti-(1.5-4.5)Zr-(0.5-5.5)Sn-(1.5-4.4)Mo- (23.5-26.5)Nb (TLM2) [45]. The TLE, TLM2, and Ti-6Al-4V (control group) samples were implanted subcutaneously and intramuscularly in rabbit models, respectively. According to GB/T16886.6-1997 testing protocol, clinical observation and histological examination were conducted at post-op 1, 2, 6, 12 and 24 weeks, respectively. The results demonstrated that the testing samples did not differ to the control group in terms of cellular infiltration. However, the thickness of envelope of the experiment groups was thinner as compared with the control. The MTT and ALP testing results suggested that higher proliferation rate and alkaline phosphatase activity were obtained on the two β samples than those on the Ti-6Al-4V alloy, when cultured with osteoblasts. All the experiments proved that TLE and TLM2 demonstrated good biocompatibility. Jia *et al.* from the ninth people's hospital of Shanghai Jiao Tong University studied the cytocompatibility of a self-invented β titanium alloy by using L-929 (mouse fibroblast cells) [46]. To investigate the IL-6 and TNF-α expression, the Ti-Nb-Zr samples were cultured with macrophages (J774A.1) for 24 to 28 hours. The results exhibited that the level of cytotoxicity of Ti-Nb-Zr was very limited. The cell morphology of macrophages devoured the particle exhibited much more normal than those devoured Cr-Mo or Ti-Al-V particles. The expressions of IL-6 and TNF-α mRNA increased significantly. However, the TNF-α expression became less than that of Co-Cr and Ti-Al-V after 48 hours of culturing. Their study proved that low modulus Ti-Nb-Zr alloys performed good biocompatibility.

A lot of literatures demonstrated that as for clinical applications, the cytotoxicity of β titanium alloys was as same as the pure titanium [47, 48]. Fig. (**3**) shows typical radiographs of rabbit femurs 12 weeks following implantation with Ti6Al4V and Ti35Nb2Ta3Zr [49]. The implants were located in the medullary canal, and no fractures were observed in the operated bone. New bone tissue was observed around the implants as indicated by the arrow in Fig. (**3**). No significant differences were observed among the radiographs for the different conditions. Fig. (**4**) shows representative histological images of the middle section of the Ti35Nb2Ta3Zr alloy at 12 weeks after implantation [49, 50]. Newly formed bone tissue was clearly observed around the surface of both the alloys and good contact was observed between the bone tissue and the Ti35Nb2Ta3Zr alloy. Under fluorescence microscopy, a double line of tetracycline and calcein was clearly observed, indicating new bone formation.

In general, the new type of β titanium alloys have low elastic modulus, high specific strength, excellent cold machinability and superior wear resistance as well as good biocompatibility. It is believed that the applications of β titanium alloys in dentistry and orthopaedics can be broadened, when the biological safety concerns have been comprehensively addressed.

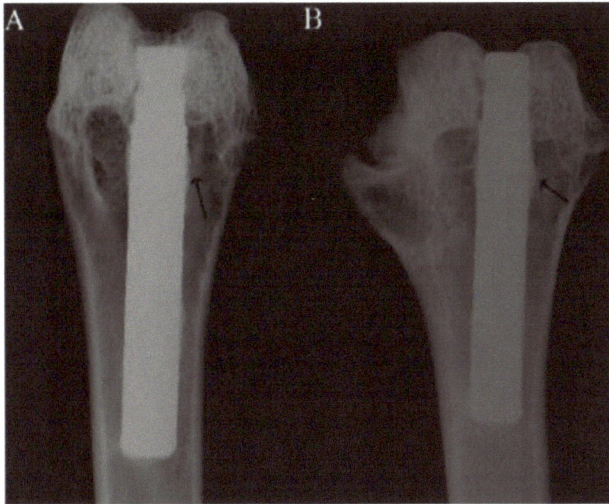

Fig. (3). Typical radiographs of Ti6Al4V and Ti35Nb2Ta3Zr alloys at 12 weeks after rod implantation. The new bone formed around the rod is indicated by arrows. (**A**) Ti6Al4V and (**B**) Ti35Nb2Ta3Zr [49].

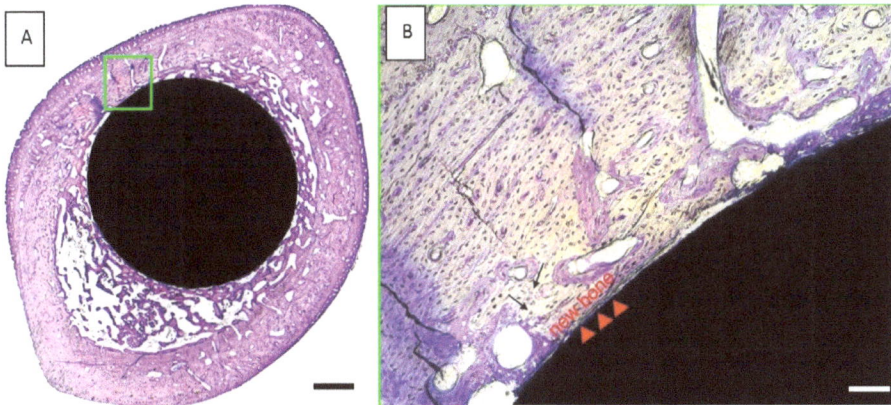

Fig. (4). Histological analysis in the middle section stained with methylene blue (Ti35Nb2Ta3Zr alloy) 12 weeks after implantation. (**A**) Integral image around the rod observed by light microscopy. Bar, 1.0 mm. (**B**) Magnification of the square area in (**A**) observed in fluorescent light micrographs. Bar, 100 m. The newly formed bone directly contacted the rod surface (triangles). The arrows indicate newly formed osteocytes [49].

PHASE TRANSFORMATIONS OF TITANIUM ALLOYS IN BIOMEDICAL APPLICATIONS

Ti-Ni Alloys

The shape memory effect (SME) is based on martensitic transformations and anti-martensitic transformations. Some of the deformed martensitic alloys would return to its pre-deformed shape of the parent phase by anti-martensitic transformation, when heated to the temperature above As (start temperature of austenite transformation). This process is called "one-way" memory effect as shown in Fig. (5a). After a series of "trainings", some materials would "remember" the shape of parent phase above As, but exhibit the shape of martensitic phase below Ms (start temperature of martensite transformation). This phenomenon is called "two-way" memory effect as shown in Fig. (5b).

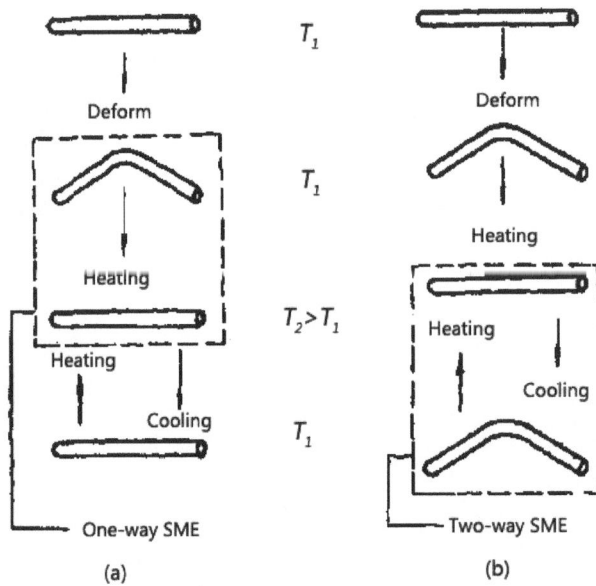

Fig. (5). A schematic illustration of shape memory effect (**a**) one-way SME; (**b**) two-way SME [51].

Ti-Nb Alloys

Fig. (6) shows the pseudo-binary diagram of titanium with the decomposition products of the β phase. As shown in the figure, with the increase of the content of β stabilizing element such as Nb, Ms decreased gradually. Therefore, the more content of Nb element is, the lower temperature of the martensite transformation will be.

Fig. (7) shows that β parent phase is in BCC structure and α' martensite is in HCP structure. Generally speaking, certain orientations of β phase and α" phase can be observed and two phases have crystal orientations expressed by equations 1.

$$[100]_\beta - [100]_{\alpha''}, [010]_\beta - \frac{1}{2}\left[01\bar{1}\right]_{\alpha''}, [001]_\beta - \frac{1}{2}[011]_{\alpha''} \qquad \textbf{(1)}$$

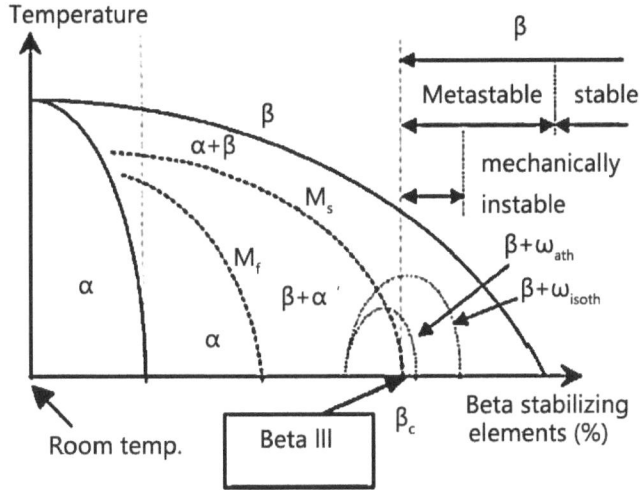

Fig. (6). Pseudo-binary diagram of titanium with the decomposition products of the β phase [52].

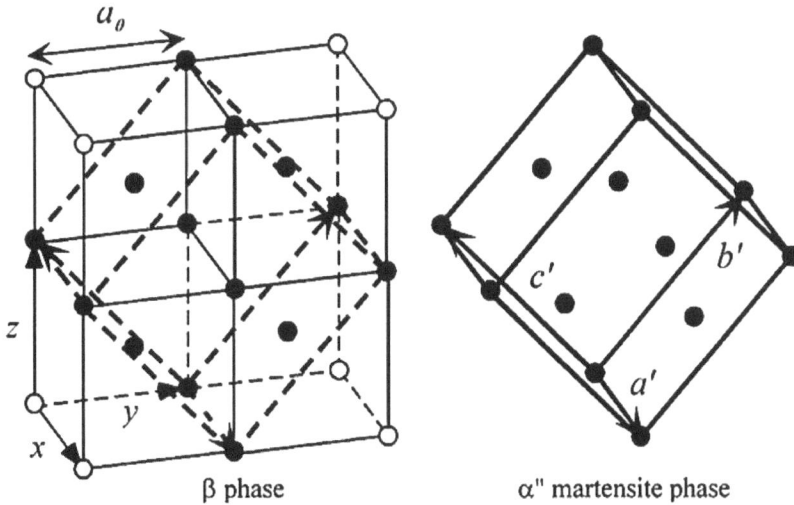

Fig. (7). A schematic illustration exhibiting lattice correspondence between β phase and α" phase.

Fig. (8) shows the sketch map of drape-shaped martensite surface caused by macroscopical shear deformation. The macroscopical structure of fracture surface changed from XY plane to XZ plane, which was attributed to the strain of lattices (R) caused by tensile stress.

Current Applications of Phase Transformation

To date, the SME of NiTi has been applied to several clinical applications. One of the best known is the Mitek suture anchor used in orthopedic surgery to attach tendons, ligaments, and other soft tissues to bones. The anchor is inserted into a small hole drilled in the bone, and expands after exposure to body temperature, forming a tight and secure anchoring point for sutures. Another application is bone shaft healing using bone plates with SME. Ohnishi (described by Funakubo *et al.* [53]) experimented with NiTi bone plates, which could be shortened if they were completely fixed between the fractured bones and heated a few degrees above body temperature. In this way, a compressive force could be applied to the ends of the bone to promote bone shaft healing.

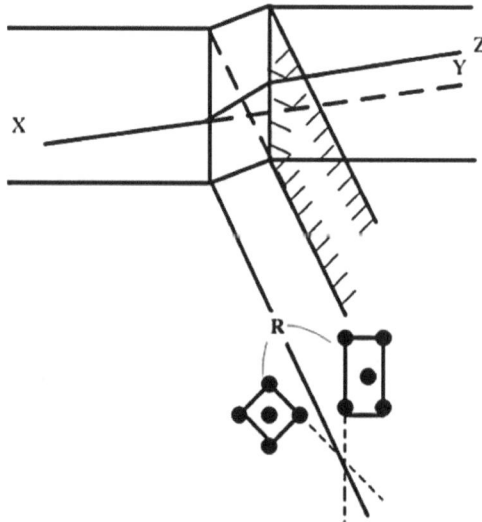

Fig. (8). Sketch map of drape-shaped martensite surface caused by macroscopical shear deformation. R: the strain of lattices [39].

One potential clinical application of the SME is to use NiTi alloy to correct scoliosis, and this has been the subject of investigations by several groups [54 - 58]. Sanders and his group used a goat scoliosis model, and then heated implanted NiTi alloy until a straight spine was obtained in the lateral view [57]. The results showed that correction of the scoliotic deformity was possible, but several drawbacks were encountered. One of the major problems was the damage of surrounding tissues, especially the spinal cord, when the rod was being heated during surgery. The temperature required to complete the shape change might be higher than expected, as the transition temperature changes when the rod is loaded [59]. The transition temperature is likely to be stress-dependent and therefore extremely difficult to control and to predict under such situations [56].

The property of superelasticity has been used in medical applications, and one of the most successful applications is the use of NiTi guidewires. These can be drawn down an extremely small hole but resist kinking where they hit hindrances that might permanently deform a wire made of stainless steel. The risk of patient injury due to guidewire failure is reduced, while the control and feedback in directing the wire to the target location is increased, as the manipulation input provided by the surgeon is completely transferred to the other end of the wire. Another use of this property has been found in dental surgery. NiTi wire has been used in orthodontic treatment. Stainless steel orthodontic wire was widely used in the past, but its stiffness and small elastic elongation range meant that it very quickly went into the range of plastic deformation, and there were problems producing a suitable corrective force at the time of installation. NiTi has a non-linear modulus because of its superelasticity, and therefore little force is required to elicit a large deformation. The manipulation is easier during installation, and the time of treatment and patient's discomfort are reduced [53].

CONCLUSION

The titanium alloy research for biomedicine is currently focusing on the development of low elastic modulus alloys. Among various metallic materials, low modulus β titanium alloys with "bio-metallic elements" including Nb, Ta and Zr are the highlights in the field and they have great potentials in clinical applications. The effects of thermo-mechanical parameters and solute strengthening treatment as well as super-elasticity and shape memory effect improvement are of vital importance in the further research on titanium alloys in biomedical applications. In the future, much attention should be paid on the biological evaluation of biomedical material and fatigue performance improvement. Also cross-disciplinary exchanges and cooperation between researchers of material science and medicine should be encouraged to join together to do some researches on biomedical titanium alloys, which is very important to the clinical medical titanium industrial applications.

CONFLICT OF INTEREST

The author (editor) declares no conflict of interest, financial or otherwise.

ACKNOWLEDGEMENTS

The work done by the authors was financially supported by National Science Foundation under Grant No: 81171738,51242008, Shanghai Natural Science Foundation under Grant No:12ZR1445500, New Teachers' Fund for Doctor Stations, Ministry of Education under Grant No: 20120073120007,Medical Engineering Cross Research Foundation of Shanghai Jiaotong University under Grant No. YG2011MS23.

REFERENCES

[1]   Akpalu W, Normanyo AK. Gold mining pollution and the cost of private healthcare: the case of Ghana. Ecol Econ 2017; 142: 104-12.
        [http://dx.doi.org/10.1016/j.ecolecon.2017.06.025]

[2]   Temenoff JS, Mikos AG. Biomaterials: the intersection of biology and materials science. Upper Saddle River, NJ: Pearson/Prentice Hall 2008.

[3]   Wang LQ, Lu WJ, Qin JN, Zhang F, Zhang D. Microstructure and mechanical properties of cold-rolled TiNbTaZr biomedical [beta] titanium alloy. Mater Sci Eng A 2008; 490: 421-6.
        [http://dx.doi.org/10.1016/j.msea.2008.03.003]

[4]   Long M, Rack HJ. Titanium alloys in total joint replacement-a materials science perspective. Biomaterials 1998; 19(18): 1621-39.
        [http://dx.doi.org/10.1016/S0142-9612(97)00146-4] [PMID: 9839998]

[5]   Sumner DR, Galante JO. Determinants of stress shielding: design *versus* materials *versus* interface. Clin Orthop Relat Res 1992; (274): 202-12.
        [PMID: 1729005]

[6]   Saito T, Furuta T, Hwang JH, *et al.* Multifunctional alloys obtained *via* a dislocation-free plastic deformation mechanism. Science 2003; 300(5618): 464-7.
        [http://dx.doi.org/10.1126/science.1081957] [PMID: 12702870]

[7]   Kuroda D, Niinomi M, Morinaga M, Kato Y, Yashiro T. Design and mechanical properties of new β type titanium alloys for implant materials. Mater Sci Eng A 1998; 243: 244-9.
        [http://dx.doi.org/10.1016/S0921-5093(97)00808-3]

[8]   Black J. Biological Performance of Materials. M Dekker 1992.

[9]   Williams DF. Biocompatibility in clinical practice. CRC Press 1982.

[10]   Van NR. Titanium: the implant material of today. J Mater Sci 1987; 22: 3801-11.
        [http://dx.doi.org/10.1007/BF01133326]

[11]   Piao M, Miyazaki SC, Otsuka K, Nishida N. Effects of Nb addition on the microstructure of Ti-Ni alloys. Mater Trans 1992; 33: 337-45.
        [http://dx.doi.org/10.2320/matertrans1989.33.337]

[12]   Park JB, Lakes RS. Biomaterials-an introduction. 2nd ed., New York: Plenum Press 1992.
        [http://dx.doi.org/10.1007/978-1-4757-2156-0_1]

[13]   Leventhal GS. Titanium, a metal for surgery. J Bone Joint Surg Am 1951; 33-A(2): 473-4.
        [http://dx.doi.org/10.2106/00004623-195133020-00021] [PMID: 14824196]

[14]   Brånemark PI, Engstrand P, Ohrnell LO, *et al.* Brånemark Novum: a new treatment concept for rehabilitation of the edentulous mandible. Preliminary results from a prospective clinical follow-up study. Clin Implant Dent Relat Res 1999; 1(1): 2-16.
        [http://dx.doi.org/10.1111/j.1708-8208.1999.tb00086.x] [PMID: 11359307]

[15]   Boyer R, Welsch G, Collings EW. Materials properties handbook: Titanium alloys. ASM Int 1993.

[16]   Steinemann SG. Corrosion of titanium and titanium alloys for surgical implants. Tire Sci Technol 1984; 2: 1373-9.

[17]   Wang K. The use of titanium for medical applications in the USA. Mater Sci Eng A 1996; 213: 134-7.
        [http://dx.doi.org/10.1016/0921-5093(96)10243-4]

[18]   Pilehva F, Zarei-Hanzaki A, Ghambari M, Abed HR. Flow behavior modeling of a Ti-6Al-7Nb biomedical alloy during manufacturing at elevated temperatures. Mater Des 2013; 51: 457-65.
        [http://dx.doi.org/10.1016/j.matdes.2013.04.046]

[19]   Gordin DM, Ion R, Vasilescu C, Drob SI, Cimpean A, Gloriant T. Potentiality of the "Gum Metal" titanium-based alloy for biomedical applications. Mater Sci Eng C 2014; 44: 362-70.

[http://dx.doi.org/10.1016/j.msec.2014.08.003] [PMID: 25280716]

[20]   Buly RL, Huo MH, Salvati E, Brien W, Bansal M. Titanium wear debris in failed cemented total hip arthroplasty. An analysis of 71 cases. J Arthroplasty 1992; 7(3): 315-23.
[http://dx.doi.org/10.1016/0883-5403(92)90056-V] [PMID: 1402950]

[21]   Song X, Wang L, Niinomi M, Nakai M, Liu Y. Fatigue characteristics of a biomedical β-type titanium alloy with titanium boride. Mater Sci Eng A 2015; 640: 154-64.
[http://dx.doi.org/10.1016/j.msea.2015.05.078]

[22]   Zhang QH, Cossey A, Tong J. Stress shielding in periprosthetic bone following a total knee replacement: Effects of implant material, design and alignment. Med Eng Phys 2016; 38(12): 1481-8.
[http://dx.doi.org/10.1016/j.medengphy.2016.09.018] [PMID: 27745873]

[23]   Ye B, Dunand DC. Titanium foams produced by solid-state replication of NaCl powders. Mater Sci Eng A 2010; 528: 691-7.
[http://dx.doi.org/10.1016/j.msea.2010.09.054]

[24]   Buehler WJ, Gilfrich JV, Wiley RC. Effect of low-temperature phase changes on the mechanical properties of alloys near composition TiNi. J Appl Phys 1963; 34: 1475-7.
[http://dx.doi.org/10.1063/1.1729603]

[25]   Wayman CM, Shimizu K. The shape memory ('Marmem') effect in alloys. Met Sci J 1972; 6: 175-83.
[http://dx.doi.org/10.1179/030634572790446028]

[26]   Laheurte P, Eberhardt A, Philippe MJ. Influence of the microstructure on the pseudoelasticity of a metastable beta titanium alloy. Mater Sci Eng A 2005; 396: 223-30.
[http://dx.doi.org/10.1016/j.msea.2005.01.022]

[27]   Zhou YL, Niinomi M, Akahori T. Decomposition of martensite α″ during aging treatments and resulting mechanical properties of Ti-Ta alloys. Mater Sci Eng A 2004; 384: 92-101.
[http://dx.doi.org/10.1016/j.msea.2004.05.084]

[28]   Yang GJ, Zhang T. Phase transformation and mechanical properties of the Ti50Zr30Nb10Ta10 alloy with low modulus and biocompatible. J Alloys Compd 2005; 392: 291-4.
[http://dx.doi.org/10.1016/j.jallcom.2004.08.099]

[29]   Yu SY, Scully JR. Corrosion and passivity of Ti-13% Nb-13% Zr in comparison to other biomedical implant alloys. Corros J 1997; 53: 965-76.
[http://dx.doi.org/10.5006/1.3290281]

[30]   Saji VS, Choe HC, Brantley WA. An electrochemical study on self-ordered nanoporous and nanotubular oxide on Ti-35Nb-5Ta-7Zr alloy for biomedical applications. Acta Biomater 2009; 5(6): 2303-10.
[http://dx.doi.org/10.1016/j.actbio.2009.02.017] [PMID: 19289307]

[31]   Kuroda D, Niinomi M, Morinaga M, Katod Y, Yashirod T. Design and mechanical properties of new β type titanium alloys for implant materials. Mater Sci Eng A 1998; 243: 244-9.
[http://dx.doi.org/10.1016/S0921-5093(97)00808-3]

[32]   Nag S, Banerjee R, Fraser HL. Microstructural evolution and strengthening mechanisms in Ti-Nb--r-Ta, Ti-Mo-Zr-Fe and Ti-15Mo biocompatible alloys. Mater Sci Eng A 2005; 25: 357-62.
[http://dx.doi.org/10.1016/j.msec.2004.12.013]

[33]   Kim HY, Satoru H, Kim JI, Hosoda H, Miyazaki S. Mechanical properties and shape memory behavior of Ti-Nb alloys. Mater Trans 2004; 45: 2443-8.
[http://dx.doi.org/10.2320/matertrans.45.2443]

[34]   Niinomi M, Akahori T, Katsura S, Yamauchi K, Ogawa M. Mechanical characteristics and microstructure of drawn wire of Ti-29Nb-13Ta-4.6 Zr for biomedical applications. Mater Sci Eng C 2007; 27: 145-61.
[http://dx.doi.org/10.1016/j.msec.2006.04.008]

[35]   Yu ZT, Zheng YF, Zhou L, *et al.* Shape Memory Effect and Superelastic Property of a Novel Ti-3Zr-2Sn-3Mo-15Nb Alloy. Rare Met Mater Eng 2008; 37: 1-5.
[http://dx.doi.org/10.1016/S1875-5372(09)60004-7]

[36]   Bai XF, Zhao YQ, Zeng WD, Zhang YS, Li B. Deformation mechanism and microstructure evolution of TLM Titanium alloy during cold and hot compression. Rare Met Mater Eng 2015; 44: 1827-31.
[http://dx.doi.org/10.1016/S1875-5372(15)30108-9]

[37]   Zhu JM, Wu HH, *et al.* Crystallographic analysis and phase field simulation of transformation plasticity in a multifunctional β-Ti alloy. Int J Plast 2017; 89: 110-29.
[http://dx.doi.org/10.1016/j.ijplas.2016.11.006]

[38]   Wang LQ, Lu WJ, Qin JN, Zhang F, Zhang D. Microstructure and mechanical properties of cold-rolled TiNbTaZr biomedical β titanium alloy. Mater Sci Eng A 2008; 490: 421-6.
[http://dx.doi.org/10.1016/j.msea.2008.03.003]

[39]   Wang LQ, Lu WJ, Qin JN, Zhang F, Zhang D. Influence of cold deformation on martensite transformation and mechanical properties of Ti-Nb-Ta-Zr alloy. J Alloys Compd 2009; 469: 512-8.
[http://dx.doi.org/10.1016/j.jallcom.2008.02.032]

[40]   Wang L, Lu W, Qin J, Zhang F, Zhang D. Texture and superelastic behavior of cold-rolled TiNbTaZr alloy. Mater Sci Eng A 2008; 491: 372-7.
[http://dx.doi.org/10.1016/j.msea.2008.01.018]

[41]   Wang L, Lu W, Qin J, Zhang F, Zhang D. Change in microstructures and mechanical properties of biomedical Ti-Nb-Ta-Zr system alloy through cross-rolling. Mater Trans 2008; 49: 1791-5.
[http://dx.doi.org/10.2320/matertrans.MRA2008040]

[42]   Eisenbarth E, Velten D, Müller M, Thull R, Breme J. Biocompatibility of β-stabilizing elements of titanium alloys. Biomaterials 2004; 25(26): 5705-13.
[http://dx.doi.org/10.1016/j.biomaterials.2004.01.021] [PMID: 15147816]

[43]   Niinomi M. Fatigue performance and cyto-toxicity of low rigidity titanium alloy, Ti-29Nb-13-a-4.6Zr. Biomaterials 2003; 24(16): 2673-83.
[http://dx.doi.org/10.1016/S0142-9612(03)00069-3] [PMID: 12711513]

[44]   Niinomi M, Akahori T, Takeuchi T, Katsura S, Fukui H, Toda H. Mechanical properties and cyto-toxicity of new beta type titanium alloy with low melting points for dental applications. Mater Sci Eng C 2005; 25: 417-25.
[http://dx.doi.org/10.1016/j.msec.2005.01.024]

[45]   Zhang MH, Yu ZT, Zhou L. Rare Met Mater Eng 2007; 36: 1815-9.

[46]   Jia QW, Ning CQ, Dind DY. Bate titanium alloys for orthopedic implants:fabrication,X-ray diffraction analysis,and biomechanics. Ortho J China 2008; 16: 430-4.

[47]   Miura K, Yamada N, Hanada S, Jung TK, Itoi E. The bone tissue compatibility of a new Ti-Nb-Sn alloy with a low Young's modulus. Acta Biomater 2011; 7(5): 2320-6.
[http://dx.doi.org/10.1016/j.actbio.2011.02.008] [PMID: 21316491]

[48]   Hu Y, Cai K, Luo Z, *et al.* Regulation of the differentiation of mesenchymal stem cells *in vitro* and osteogenesis *in vivo* by microenvironmental modification of titanium alloy surfaces. Biomaterials 2012; 33(13): 3515-28.
[http://dx.doi.org/10.1016/j.biomaterials.2012.01.040] [PMID: 22333987]

[49]   Guo Y, Chen D, Cheng M, Lu W, Wang L, Zhang X. The bone tissue compatibility of a new Ti35Nb2Ta3Zr alloy with a low Young's modulus. Int J Mol Med 2013; 31(3): 689-97.
[http://dx.doi.org/10.3892/ijmm.2013.1249] [PMID: 23338484]

[50]   Guo Y, Chen D, Lu W, Jia Y, Wang L, Zhang X. Corrosion resistance and *in vitro* response of a novel Ti35Nb2Ta3Zr alloy with a low Young's modulus. Biomed Mater 2013; 8(5): 055004.
[http://dx.doi.org/10.1088/1748-6041/8/5/055004] [PMID: 24002775]

[51]   Xu ZY. Martensite transformation and martensite. Sci Press 1999.

[52]   Laheurte P, Eberhardt A, Philippe MJ. Influence of the microstructure on the pseudoelasticity of a metastable beta titanium alloy. Mater Sci Eng A 2005; 396: 223-30.
[http://dx.doi.org/10.1016/j.msea.2005.01.022]

[53]   Funakubo H. Application of shape memory alloys.Shape Memory Alloy. Gordon and Breach Science 1987; pp. 227-69.

[54]   Lu SB, Wang JF, Guo JF. Treatment of scoliosis with a shape-memory alloy rod. Zhonghua Wai Ke Za Zhi 1986; 24(3): 129-132, 187.
[PMID: 3757641]

[55]   Matsumoto K, Tajima N, Kuwahara S. Correction of scoliosis with shape-memory alloy. Nihon Seikeigeka Gakkai zasshi 1993; 67: 267-74.

[56]   Schmerling MA, Wilkov MA, Sanders AE, Woosley JE. Using the shape recovery of nitinol in the Harrington rod treatment of scoliosis. J Biomed Mater Res 1976; 10(6): 879-92.
[http://dx.doi.org/10.1002/jbm.820100607] [PMID: 993225]

[57]   Sanders JO, Sanders AE, More R, Ashman RB. A preliminary investigation of shape memory alloys in the surgical correction of scoliosis. Spine 1993; 18(12): 1640-6.
[http://dx.doi.org/10.1097/00007632-199309000-00012] [PMID: 8235844]

[58]   Veldhuizen AG, Sanders MM, Cool JC. A scoliosis correction device based on memory metal. Med Eng Phys 1997; 19(2): 171-9.
[http://dx.doi.org/10.1016/S1350-4533(96)00049-5] [PMID: 9203152]

[59]   Liu Y, Galvin SP. Criteria for pseudoelasticity in near-equiatomic NiTi shape memory alloys. Acta Mater 1997; 45: 4431-9.
[http://dx.doi.org/10.1016/S1359-6454(97)00144-4]

Microstructure and Mechanical Properties of TiNbTaZr Titanium Alloy

Liqiang Wang[1,*], **Xueting Wang**[1] and **Lai-Chang Zhang**[2]

[1] *State Key Laboratory of Metal Matrix Composites, Shanghai Jiao Tong University, No. 800 Dongchuan Road, Shanghai 200240, P.R. China*

[2] *School of Engineering, Edith Cowan University, 270 Joondalup Drive, Joondalup, Perth, WA, 6027, Australia*

Abstract: Currently, Ti-Mo, Ti-Nb, Ti-Ta and Ti-Zr-based β-titanium alloys have been widely studied. Compared with other titanium alloys, these alloys can achieve lower elastic modulus and higher strength. Since Ti-Nb alloy has low elastic modulus and good shape memory effect, it is the most promising medical titanium alloy to develop and utilize. Beta titanium alloys with the elements of Nb,Ta,Zr are being studied as the most important biomedical materials. This chapter presents the microstructure and mechanical properties of TiNbTaZr titanium alloy. The results show that the TiNbTaZr β titanium alloy with lower elastic modulus and non-toxic alloying elements has much more important application in biomedical field. The Ti35Nb2Ta3Zr β titanium alloy is studied and the alloy has the complex properties of lower elastic modulus, high strength, high elongation and excellent shape memory effect. Compared with direct rolling, cross rolling is beneficial to the isotropic of the microstructure and mechanical properties.

Keywords: β titanium alloy, Elastic modulus, Mechanical Properties, Microstructure.

INTRODUCTION

In today's field of medical materials, the manufacture of devices increasingly demands for materials. These devices are mainly artificial joints, bones, internal and external fixation devices in orthopedic, Others include dentures, fillings, implants, orthopedic wires and a variety of adjuvant therapy devices in dentistry. Metal material is firstly applied, and most widely used in the current clinical application. Initially, the metal material used in clinical was stainless steel with good corrosion resistance performance, among which the most commonly used

* **Corresponding author Liqiang Wang:** State Key Laboratory of Metal Matrix Composites, School of Materials Science and Engineering, Shanghai Jiao Tong University, No. 800 Dongchuan Road, Shanghai 200240, P.R. China; Tel: 8602134202641; Fax: 8602134202749; E-mail: wang_liqiang@sjtu.edu.cn

was 316L austenitic stainless steel. Co-Cr alloy was developed later. These series of alloys have good corrosion resistance and strength in the biological environment, thus are widely used in biomedical fields [1]. In early 1940s, Bothe *et al.* [2] published articles on a variety of reactions between the metal implant and bones, after titanium was introduced in the biomedical field. Bothe's group had several metal implanted in rats' femur, including titanium, stainless steel and Co-Cr alloy, and found that there were no adverse reactions between the titanium and the bone. Possessing high biocompatibility, low density, low modulus, high strength, corrosion-resistance in body fluids, and other advantages, titanium alloys are widely used in medical field. Fig. (**1**) shows the application of medical material in the joints connection parts of human body [3].

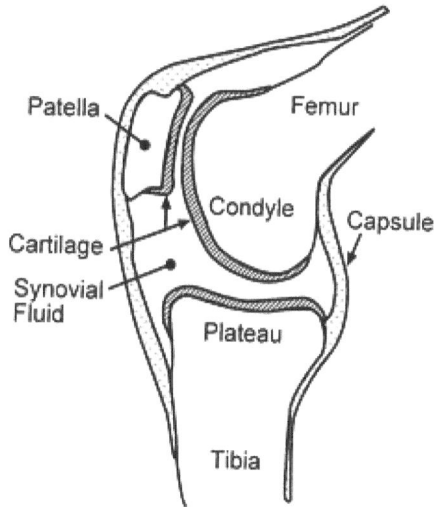

Fig. (1). Medical materials in the joints connecting the body parts of the application [3].

TiNi Shape Memory Alloy

As early as 1963, Buehler *et al.*, the U.S. Navy Ordnance Laboratory, have already found shape memory effect in near-equiatomic Ti-Ni alloy, and launched commercial TiNinol alloy projects. In 1969, the researchers in the United States have produced shape memory alloy in batches, and put it into use. Then, the new titanium alloys with superelasticity and shape memory effects have been widely studied and used in aerospace, offshore oil, cable communications and other fields [4, 5]. From 1978, China has carried out a wide range of corrosion resistant experiments on biological testing for TiNi shape memory alloy and its products. The Results showed that TiNi alloys have good biocompatibility, as well as brilliant corrosion resistance in various physiological solution or medium. After the implantation of its various products, neither rejection nor inflammation was observed in the body. So far, shape memory alloys have been used widely in clinical practice, such as artificial joints, dental orthodontics, and root canal setbacks. Due to the significant shape memory effect, pseudo-elasticity, corrosion resistance and biocompatibility, biomedical TiNb shape memory alloys, have been developed widely and applied in recent years. Currently, many researchers made a lot of effort to study the shape memory effect of Ni and Ti-rich TiNi alloy [6 - 8], the ultimate goal is to further improve the alloy's shape memory effect. The researches mainly focus on the following areas:

- Cold deformation, thermo-mechanical processing and aging heat treatment on the phase transition and physical properties of Ni-rich TiNi alloy.
- The impacts of adding alloying elements on phase transformation and physical properties of TiNi alloy.
- Surface treatment improves corrosion resistance and biocompatibility of TiNi alloy. When replaced Ni in the TiNi alloy with Au, Pd, Pt, Zr, Hf and other elements, we can further improve the used temperature of the alloy, consequently expanding its applicable areas.

Biomedical β Titanium Alloy

According to the type of titanium alloys (α, α + β, β), biomedical development and application of titanium alloys can be generally divided into three generations. Pure titanium and Ti-6Al-4V were developed in the first generation. Nowadays, because of the properties of good biocompatibility, corrosion resistance and mechanical properties, Ti-6Al-4V is the most widely used surgical implant material. The second generation of titanium alloys is α + β type alloy, and represented by Ti-5Al-2.5Fe and Ti-6Al-7Nb. The alloys of Ti-6Al-7Nb and Ti-5Al-2.5Fe, which are two kinds of α + β titanium alloy without toxic element of V, have the similar mechanical properties as Ti-6Al-4V ELI alloy, and also widely used in surgical fields. Compared to the elastic modulus of human bones, which is around 10-30GPa, such titanium alloys have larger elastic modulus. Thus, after being implanted in the body, the alloys bear much more forces than bones, therefore generate "stress shielding" phenomenon, which could harm the recovery of body. Therefore, the research goal of the third generation titanium alloy is to achieve lower elastic modulus. In the 1990s, Ti-Mo system β-type titanium alloys have been extensively studied as medical materials, such as Ti-12Mo-6Zr-2Fe, Ti-15Mo-5Zr-3Al, Ti-15Mo-3Nb-0.3O (21SRx), *etc.* [9, 10]. These type of alloys have higher tensile strength, higher fracture toughness, better wear resistance. However, the elastic modulus of these alloys is still higher than that of the bone. Recently, designing and developing the β-titanium alloys with lower elastic modulus has become the key scientific issue on the development of medical titanium materials. In the research of biomedical titanium alloys, it is found that some metal ions react with human organs, and a potential toxic effect appears after long-term *in vivo* implantation. Therefore, reducing and avoiding the toxic effects from alloying elements is particularly important in the development of biomedical titanium alloys. Steinmann's [11] studies have shown that, V, Ni, and Co were found to be bio-toxic elements. Meanwhile, Kawahara reported that Al, Fe had toxic effects on human body. However, a large number of experiments have proved that, the elements of Nb, Ta, Zr, and Sn had good biocompatibility and low toxicity to the human body. Thus, the β titanium alloys with low elastic modulus containing the elements of Nb, Ta, Zr and Sn would have greater application potential.

Although there are no detailed reports about the toxicity of TiNi alloy, it is likely that long-time implanted in the body, the element of Ni would generate toxicity to human being, which limits its long-term use in body. In 1971, Baker discovered that in certain conditions, Ti-35wt% Nb β titanium alloy had shape memory effect and super elasticity. The appearance of shape memory effect in β-titanium alloys made a great potential for the development of β titanium alloy in the biomedical field. Meanwhile, apart from Ti-Nb alloy, in other alloys, such as Ti-V-based titanium alloys [12], Ti-Mo based titanium alloys [13, 14], shape memory effect was also found. The emergence of martensitic in such alloys also confirms the existence of shape memory and super elasticity effects in β titanium alloys.

Currently, Ti-Mo, Ti-Nb, Ti-Ta and Ti-Zr-based β-titanium alloys have been widely studied [15 - 23]. Compared with other titanium alloys, these alloys can achieve lower elastic modulus and higher strength. Since Ti-Nb alloy has low elastic modulus and good shape memory effect, it is the most promising medical titanium alloy to develop and utilize [24 - 27].

Some of the Ti-Nb based meta-stable titanium alloys have been applied, such as Ti-29Nb-13Ta-4.6Zr [28], Ti- 39Nb-5.1Ta-7.1Zr (TNTZ) and Ti-34Nb-9Zr-8Ta [16]. These alloys have high strength, low modulus. However, the elastic modulus of these alloys are still very high (55-89GPa) when compared with the elastic modulus of human bones. Designed and developed independently by Northwest Institute for Non-ferrous Metal Research (NIN) and other two units, two kinds of "lower cost" near β-titanium alloys TLM (Ti-Zr-Sn-Mo-Nb) and TLE (Ti-Zr-Mo-Nb), have good biomedical and mechanical compatibility, as well as an overall performance comparable to the medical titanium reported globally at present, for example, Ti-13Nb-13Zr, *et al.*

Recently the Kim group [22, 23] has studied mechanical properties of Ti-Nb alloy, and found shape memory effect in Ti-(22-25) at% Nb and Ti-(25.5-27) at%

Nb alloy. While some other scholars reported the relationship between Nb content and the changes of elastic modulus. Ping [29] studied the shape memory effect of Ti-30Nb-3Pd alloy by analyzing effects of martensitic transformation in the alloy to its performance. Since Pd has excellent corrosion resistance, the corrosion rate of such alloys is minimal in saline solution.

Recently, Japan Aichi Steel Corporation has developed a new titanium alloy, which is both "flexible" and "strength". Being a Ti-Nb-Mo β titanium alloy, the newly developed alloy overmatches alloys of other types (α alloys, α + β alloys) for its excellent strength and processing properties, such as low elastic modulus and other characteristics. By adding the 3rd, the 4th and other elements, the researchers managed to find out the best composition for the alloy, and succeeded in improving the mechanical properties at the same time. Its elastic modulus was about 45GPa, nearly half of the commonly-used Ti-6AI-4V, while at a similar level with magnesium alloy. Increased by 20%, its tensile strength reached 1250MPa. Compared with that of general metal materials, the elastic deformation of this alloy increased by nearly 2% [30]. In addition, such types of new material with features of super-elastic, super plastic, super elastic constant, low thermal expansion *et al.*, such as β-titanium alloy, with Ti3 (Ni, Ta, V)+ (Zr, Hf)+ O as components, have been applied in the same aspects as rubber metal. The already developed alloy compositions include Ti-12Ta-9Nb-3V-6Zr-O and Ti-23Nb-0.7Ta-2Zr-O, *etc.* Processed at room temperature, their elastic modulus can be raised up to 40GPa, strength, greater than 1000MPa, which indicates a class of β-titanium alloy with marked performances.

Elements Design for New Generation of β Titanium Alloy

Currently, the study of bio-titanium alloy with low elastic modulus focus on two factors: First, the toxicity of alloying elements; Second, the reduction of the elastic modulus of the alloys. Kawahara proved that V and Fe are toxic to human beings, the element of Al could also do harm to human body by reacting with tissue cells. However, the elements of Nb, Ta, Zr, Mo, and Sn have good biocompatibility and low toxicity. As showed in Figs. (**2a-b**), considered as low toxicity elements, Nb, Ta, Zr, Mo and Sn are being chosen to design a new type of β-titanium alloys.

When mentioned the elastic modulus of titanium alloys, there is few theory to explain the high strength combined with low modulus in biological titanium alloys. In a series of calculation and experiments concerning the combination of strength and elastic modulus of β titanium alloys such as Ti-M (M = V, Cr, Mn, Fe, Zr, Nb, Mo, W, Ta) conducted by Song *et al.* [31], it was found that low elastic modulus and high strength can be obtained in the β titanium alloys with elements of Nb, Mo, Zr, Ta, *etc*. It is obvious that with a good biocompatibility, Nb, Ta, Zr, Mo, and Sn have great potential in the design of novel low-modulus biomedical β titanium alloys.

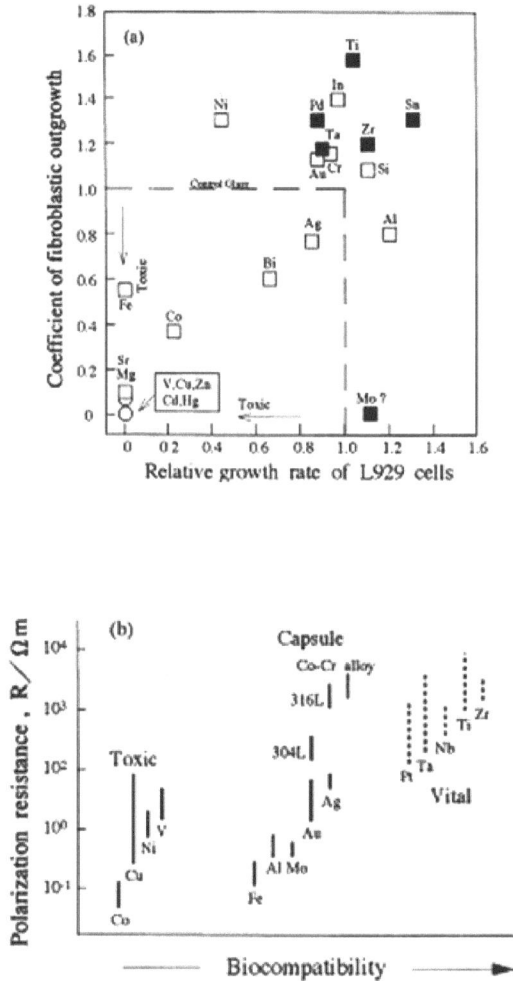

Fig. (2). Biomedical safety of metals: (a) toxicity of pure metals, (b) relationship between polarization resistance and biocompatibility of Co-Cr alloy, stainless steels and pure metals.

The Mechanical Properties of β Titanium Alloys

As load-bearing material in the body, biomedical titanium alloys should have certain intensity to avoid fracture in the body. According to the demands of biomechanics compatibility, biomedical titanium alloys should have both the properties of an elastic modulus similar to natural bone and the maximum "permissible stress". Therefore, biomedical titanium alloys should combine low elastic modulus with a relatively high strength.

After the metal implants are implanted into organisms, because of body's movement, the tissue is bound to withstand a variety of forces, tension, compression, and bending included. When these forces are mostly undertook by implants, the bone and muscle around the implant will not get a normal exercise, which leads to decreasing thickness of bone tissue, osteoporosis, muscle atrophy, *etc.*, a phenomenon known as "stress shielding". The biomechanical incompatibility appears, either too high or too low. Metal with low elastic modulus will deform greatly under certain stress, which would lose the function of fixing and supporting. Thus, the general hope is that the elastic modulus of metal is as close as that of human bones or slightly higher than the elastic modulus of human bones. Generally speaking, the value of elastic modulus of biomedical titanium alloys is required to be ranged in 60-90GPa. The elastic modulus of titanium alloy is about 100-120GPa, which is the closest to bones' elastic modulus among all biomedical metal materials. The elastic modulus of β titanium alloys is generally lower than that of α+β titanium alloys. The elastic modulus of some β titanium alloys has reached the level of carbon fiber composite material (65GPa). Table **1** lists the mechanical of some commonly used titanium implants.

Table 1. Mechanical properties of typical medical titanium alloys.

Material	Elastic modulus / GPa	Yield strength / MPa	Tensile strength / MPa
cpTi	105	692	785
Ti-6Al-4V	110	850-900	960-970
Ti-6Al-7Nb	105	921	1024
Ti-5Al-2.5Fe	110	914	1033
Ti-12Mo-6Zr-2Fe	74-85	1000-1060	1060-1100
Ti-15Mo-3Nb-0.3O	82	1020	1020
Ti-15Mo-5Zr-3Al	75	870-968	882-975
Ti-Zr	/	/	900
Ti-13Nb-13Zr	79	900	1030
Ti-35Nb-5Ta-7Zr	55	530	590
Ti-35Nb-5Ta-7Zr-04O	66	976	1010
316L stainless steel	200	170-750	465-950

It is obvious from Table **1** that commonly used titanium alloys have high strength. The elastic modulus of α+β titanium alloys reaches the half value of 316L stainless steel, while as for two-phase titanium alloys, Ti-6Al-4V for instance, hold an elastic modulus of 4-10 times to the value of human bones. Therefore, the design and development of β titanium alloy with lower elastic modulus is of vital importance in the development of medical titanium alloy materials.

MATERIAL PREPARATION AND EXPERIMENTAL METHOD

Material Preparation

The main raw materials used in the experiment contain: high pure sponge of titanium (99%), zirconium sponge (purity > 98.8%), niobium (purity > 98.5%), tantalum (purity > 99%). They are prepared adopting vacuum consumable arc melting furnace. The ingot is re-melted at least three times to ensure compositional homogeneity. The melted ingot is hot forged at 950 °C. Fig. (**3**) shows the whole process of the alloy preparation. The ingots are treated by direct and cross cold rolling. The direction of the rolling has not changed during the rolling process of direct rolling; As for the cross rolling, the rolling direction is changed by 90° for each pass [32 - 35].

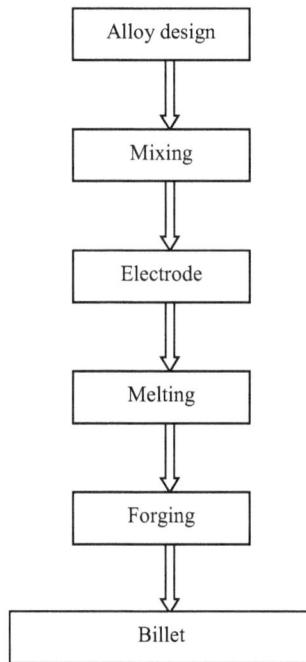

```
┌──────────────────┐
│   Alloy design   │
└──────────────────┘
          │
          ▼
┌──────────────────┐
│      Mixing      │
└──────────────────┘
          │
          ▼
┌──────────────────┐
│    Electrode     │
└──────────────────┘
          │
          ▼
┌──────────────────┐
│     Melting      │
└──────────────────┘
          │
          ▼
┌──────────────────┐
│     Forging      │
└──────────────────┘
          │
          ▼
┌──────────────────────────────┐
│            Billet            │
└──────────────────────────────┘
```

Fig. (3). Flow process chart for the preparation of the alloy.

Experiment Equipments and Analysis Methods

The room temperature mechanical and super-elastic properties are tested using plate tensile specimens. All specimens are prepared from the cold-rolled sheets by wire cutting, grinding, and polishing with metallographic sand paper, in order to eliminate surface defects. The detailed tensile specimen is shown in Fig. (**4**) (unit: mm). The gage length and thickness of the tensile specimen are 20mm and 0.3-

1.5mm, respectively. Superelastic properties are measured through cyclic loading–unloading tensile test using Zwick T1-Fr020TN materials testing machine at a strain rate of 1.5×10^{-4} s at room temperature. Specimens are stretched by 1.5%, 2.5%, 4.0%, 5.5% respectively to get superelastic stress-strain curve.

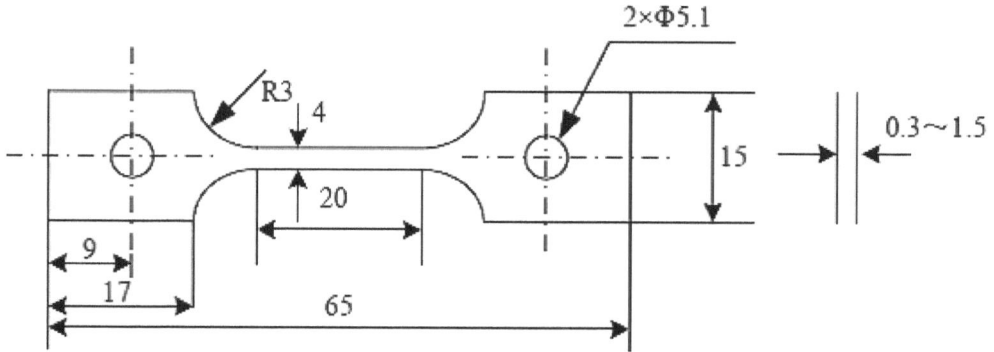

Fig. (4). Specification diagram of tensile specimens.

Calculation of β Phase Transition Point

Fig. (5) shows the Ti–Nb phase diagram. It is shown that when the Nb content is 35% (wt%), the α + β → β transition point is lower than 873K. Morinaga and others researchers have calculated the relationship between α+β →β titanium transition point and $\overline{B_0}$, $\overline{Md}$, As shown below:

$$\overline{B}_0 = 0.326\overline{Md} - 1.950 \times 10^{-4}T + 2.217 \qquad (1)$$

Calculated from the above equation, the α + β → β transition point of Ti-35N--3Zr-2Ta (wt%) is 760K. when Solution treated at 780 °C.

Martensite Start Transformation Point (Ms)

Fig. (6) reveals the DSC curve of the solution-treated Ti-35Nb-3Zr-2Ta alloy. The martensitic transformation peaks can be clearly observed in the heating and cooling process. The measured martensite start temperature (Ms) are 197.5K and 199.7K and the finished martensitic transformation temperature (Mf) are 169.9K and 170.1K. Because a large amount of β stabilizing element Nb is added, Ms is reduced under ambient.

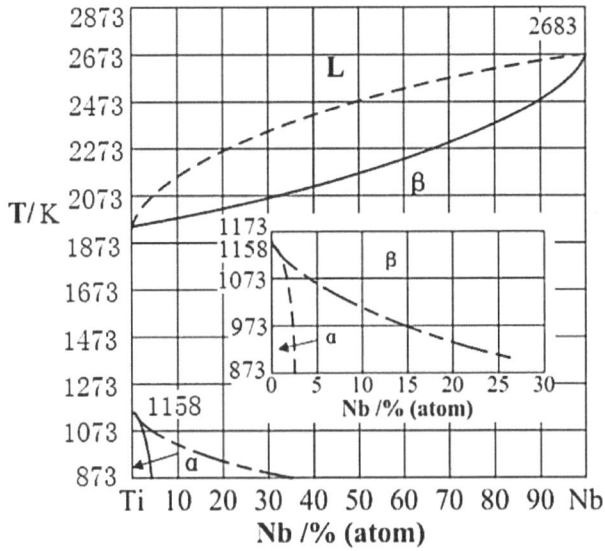

Fig. (5). Ti-Nb Phase diagram [34].

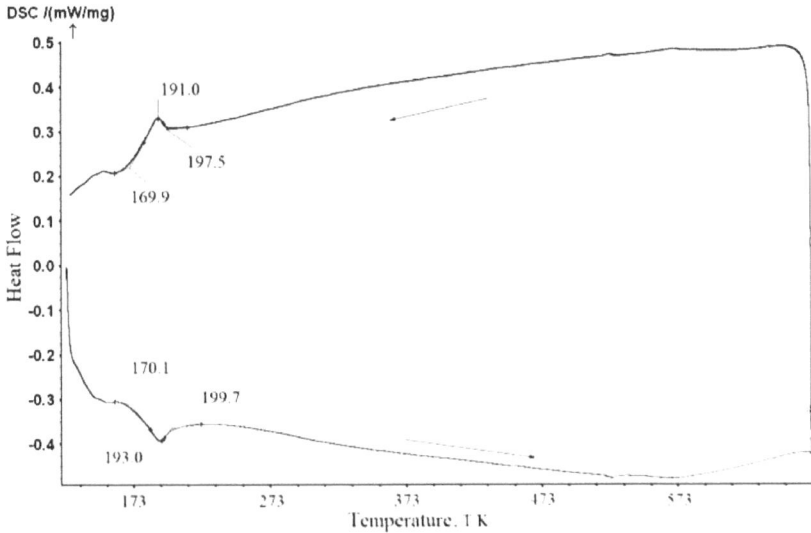

Fig. (6). DSC curve of solution-treated Ti35Nb3Zr2Ta alloy.

INFLUENCE OF DIRECT AND CROSS ROLLING ON MARTENSITE TRANSFORMATION AND MECHANICAL PROPERTIES

Specimen Preparation

The studied alloy is direct and cross rolled using two-roll mill, with the reductions of 20%, 40%, 60%, 80%, 90% and 99% respectively, and pass reduction ratio is controlled within 2% to 5%. According to the cold deformation rate, the hot-forged slabs are cut by cutting machine into rolling specimens of different thickness before rolling, and then polished by sandpapers to remove surface defects. The specimens before and after rolling with different deformation processing are shown in Fig. (7).

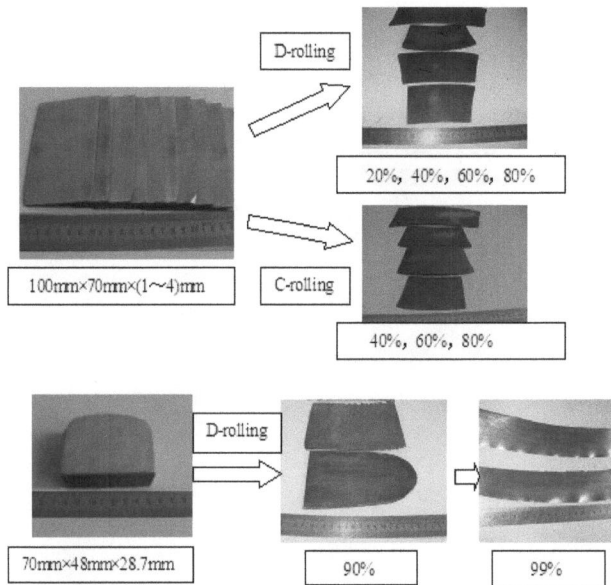

Fig. (7). Profiles of the TiNbZrTa specimens before and after rolling at different ratios of cold deformation.

Martensite Transformation in Direct Rolling

The alloy specimens are direct cold rolled by 20%, 40%, 60%, 80%, 90% and 99% after solution treatment at 780°C for 0.5 hours. The phases of specimens with different cold deformation ratio are tested by XRD. Fig. (8) for the phase analysis of the alloy after solution treatment and cold rolling. The alloys subjected to solution treatment comprise a fully retained β phase. When the alloy is rolled by 20% in thickness, β phase and stress-induced α" martensite is visible. When the deformation reduction increases to 40%, the amount of α" phase peaks increases, as well as the diffraction peak intensity ratio between α" phase to β phase. It is easy to identify that in this process the relative amount of α" martensite phase increases with the increase of cold deformation rate. When the cold deformation rate is between 60% and 80%, diffraction peak intensity of α" phase changes little, accompanied by the same trend of martensite's relative content. In the deformation process with a rate above 90%, β matrix phase appears obvious orientation. As the strain rate increases, the content of α" martensite grows gradually and becomes stable around a certain deformation rate.

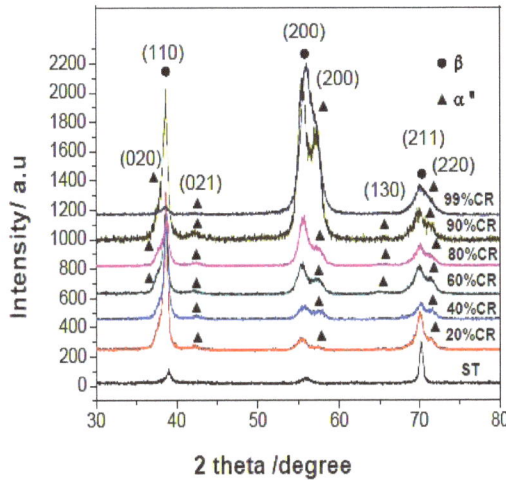

Fig. (8). X-ray diffraction profiles of the samples of solution treated (ST) and cold rolled (CR).

Martensite Morphology During Direct Rolling Process

Fig. (**9**) shows the martensitic microstructure of the specimens cross-rolled by 20%, 40%, 60%, 80% and 90%. Fig. (**9a**) shows the stress-induced martensite phase in the sample deformed by 20%, as a parallel distribution of small micro-twins. A lot of butterfly-shaped martensite is indicated in the alloy at the reduction ratio of 40%, called the butterfly-like martensite. As for TiNbZrTa β titanium alloy, when cold deformation rate is 40%, the boundary of martensite in the alloy distorts and tends to extend along the direction of deformation. Fig. (**9c**) shows morphology of martensite with the cold deformation of 60%, the "butterfly-like" martensite is stretched. With the reduction of cold rolling, much variant crossed martensite tending to rolling direction could be seen. Due to a large amount of cold deformation rate, the martensite coarsens gradually, and also exhibits obvious oriented features. Deformed at greater rate to 80%, the material shows characteristic of inter-tangled dislocation and cells-like cluster structure. Dislocations' extension and slipping contribute to the great density of dislocation in this condition. This thick martensite film is crossing each other and has mosaic morphology. Compared with 60% cold deformed martensite, this coarse plate martensite with 80% cold reduction has little trend to grow up. It is the external forces that cause interaction and woven-like extension of the plate martensite, forming deformation as an effect (Fig. **9d**). Fig. (**9e**) presents the microstructure of 90% cold deformed martensite. Its martensite texture is visible along particular rolling direction. During direct rolling, the plate martensite is sheared into small laminae with the increase of deformation ratio.

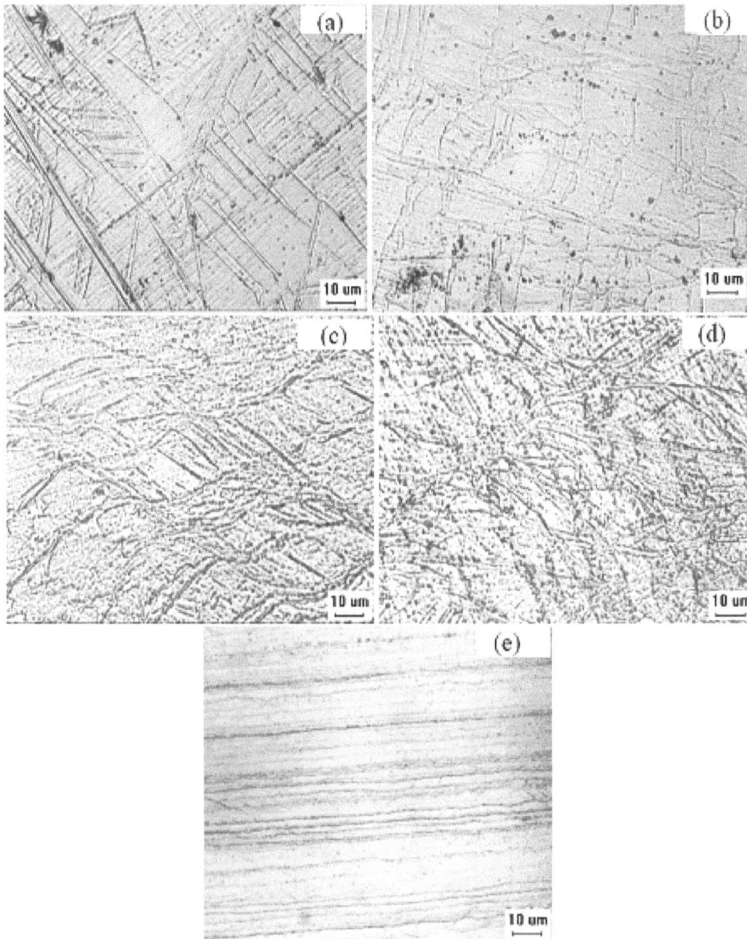

Fig. (9). Optical microscope images of plates cold rolled with the reductions of 20%(a), 40%(b),60%(c), 80%(d) and 90%(e) [34].

Fig. (**10a**) illustrates the spatial location of Ti, Nb, Zr, Ta atoms. There is flowing-water-like non-continuous lamellar structure in the specimen rolled by 60% reduction. As mentioned earlier, the sheared body-centered cubic β phase of

TiNbZrTa alloy can change to orthorhombic structure α" phase. Fig. (**10b**) indicates this micro-crystals characteristic of shear process.

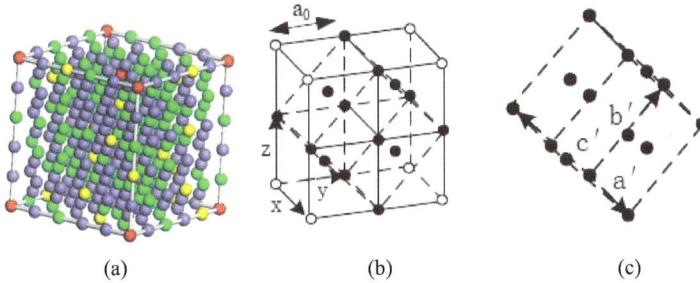

(a)   (b)   (c)

Fig. (10). The cell structure and crystallographic feature of the martensitic transformation in Ti-22.5Nb-2-r-0.7Ta (at.%) alloy.

Martensite Transformation During Cross Rolling Process

Fig. (**11**) is the XRD phase analysis of cold deformed alloy at different reductions. After the cross-rolling cold deformation, strain-induced α" phase martensite and the β phase appear in the alloy matrix. Unlike the XRD phase analysis of direct rolling, there are many martensite diffraction peaks when the reduction of cross-rolling is up to 40%. Thus it can be inferred that in cross-rolling process under the condition of relatively small deformation, much more martensite could generate. With the increase in the cold deformation ratio, martensite diffraction peaks nearly remain unchanged. In the larger deformation process, while dislocation density will increase in cold-rolled specimen, leading to the gradual stabilization of martensite.

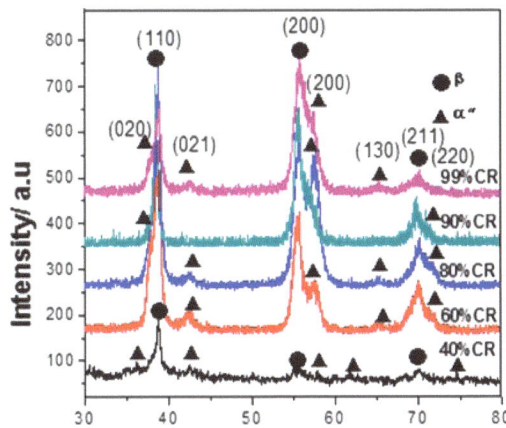

Fig. (11). X-ray diffraction profiles of the cross rolling specimens with cold reductions of 40%, 60%, 80%, 90% and 99% [35].

Martensite Morphology During Cross Rolling Process

Fig. (**12**) shows the martensitic microstructure of the specimen cross-rolled by 20%, 40%, 60% and 80%. Strain-induced martensite has a butterfly-like appearance. This martensite is similar with that of the direct-rolled specimens at same reduction. In the specimens with the reduction of 20%, there is non-oriented needle-like martensite phase. During the deformation process, strain-induced martensite appears. Meanwhile, dislocations and twins also appear. Their collision with each other provides more nucleation locations for martensite transformation. With the increase of reduction, under the effect of cross rolling shear stress, the induced martensite also demonstrates intersection behavior. Seen in Fig. (**12c**), there is flowing-water-like non-continuous lamellar structure in the specimen rolled by 60% reduction. In addition, brought by the crossed shear stress, anisotropy of variant martensite's orientation is vague. When the alloy is cross rolled by 80%, the growth of martensite can be observed clearly (Fig. **12d**).

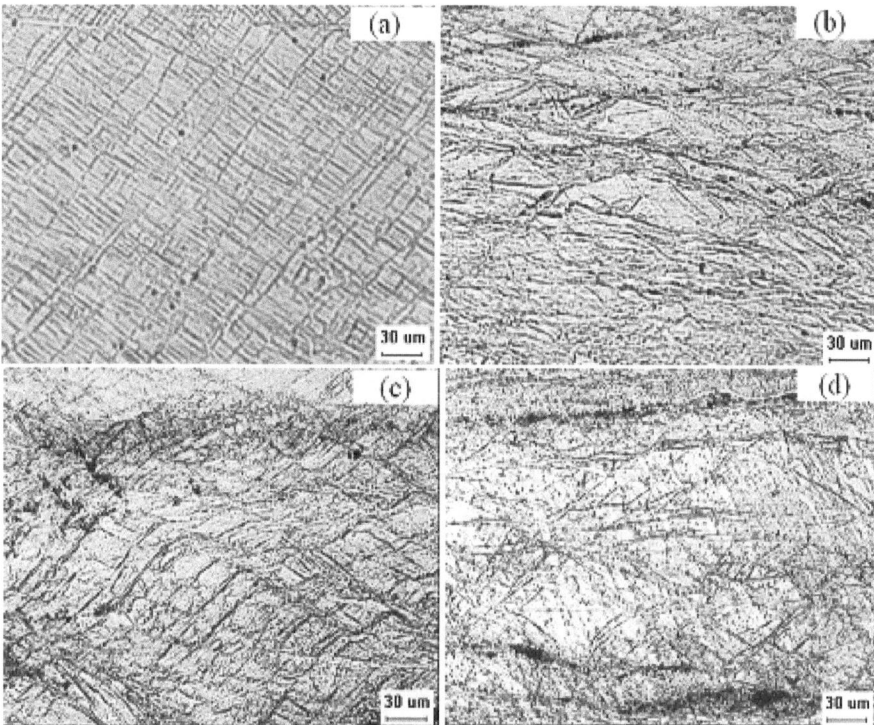

Fig. (12). Optical microscope images of plates cross rolled with the reductions of 20%(a), 40%(b), 60%(c) and 80% (d) [35].

Strain-induced Martensite Transformation

The calculation results of the amount of β phase and α" phase in direct and cross rolling process are shown in Table **2** and **3**. For direct rolling, the amount of α" phase martensite will surpass 56% when the deformation rate is 20%. With the increase of reduction under rolling stress, strain-induced martensite transformation from less-stabilized β phase to α" phase is enhanced. When the reduction adds up to 99%, the amount of martensitic transformation reaches 78.93%. As for cross rolling, martensite transformation in the deformed alloy rolled by 40% has already come to 79.63%, while the relative volumes of both β phase and α" phase have little change as reduction adds up. Compared with the direct-rolled alloy deformed by 40%, the relative volume of 99% cross-rolled alloy shows slight difference. The curve in Fig. (**13**) indicates the relationship between the cold deformation ratio and the volume of martensite. It's more favorable for the martensite transformation when processed by cross-rolling. Cross rolling is more conducive to the generation of martensite at relatively low rate of cold deformation, for the reason that the crossed shear stress in this process can make it easier for martensite transformation. Consequently, strain-induced martensite transformation arrives at a high value when the cold deformation rate is only 40%. At the same time along with martensite transformation from retained β phase to α" phase, mechanical stability of the parent phase occurs. Thus it is difficult for the stabilized β phase to continue its martensite transformation if the strain is above a certain value. It is found from the above calculations that for either TiNbZrTa alloy after direct rolling or cross rolling, the limit martensitic transformation ratio is around 79%.

Table 2. Volume Fraction of β phase and α" during direct rolling.

Deformation Rate(%)	Crystal Plane		Phase Volume Fraction(%)		Average Volume Fraction(%)	
	β phase	α" phase	β phase	α" phase	β phase	α" phase
20	(110)	(020)	41.45	58.55	43.89	56.11
	(200)	(200)	46.33	53.67		
40	(110)	(020)	38.61	61.39	36.52	63.48
	(200)	(200)	34.43	65.57		
60	(110)	(020)	26.35	73.65	27.79	72.21
	(200)	(200)	29.24	70.76		
80	(110)	(020)	21.70	78.30	27.75	72.25
	(200)	(200)	33.80	66.20		
90	(110)	(020)	25.32	74.68	26.90	73.10
	(200)	(200)	28.47	71.53		

(Table 2) contd.....

Deformation Rate(%)	Crystal Plane		Phase Volume Fraction(%)		Average Volume Fraction(%)	
	β phase	α" phase	β phase	α" phase	β phase	α" phase
99	(110)	(020)	13.54	86.44	21.07	78.93
	(200)	(200)	23.58	76.42		

Table 3. Volume fraction of β phase and α" during cross rolling.

Deformation Rate(%)	Crystal Plane		Phase Volume Fraction(%)		Average Volume Fraction(%)	
	β phase	α" phase	β phase	α" phase	β phase	α" phase
40	(110)	(020)	20.30	79.70	20.37	20.37
	(200)	(200)	20.44	79.56		
60	(110)	(020)	11.55	88.45	27.79	72.21
	(200)	(200)	30.26	69.74		
80	(110)	(020)	20.04	79.96	27.75	72.25
	(200)	(200)	25.85	74.15		
90	(110)	(020)	16.74	83.26	21.44	78.56
	(200)	(200)	26.17	73.83		
99	(110)	(020)	15.67	84.33	20.71	79.29
	(200)	(200)	25.75	74.25		

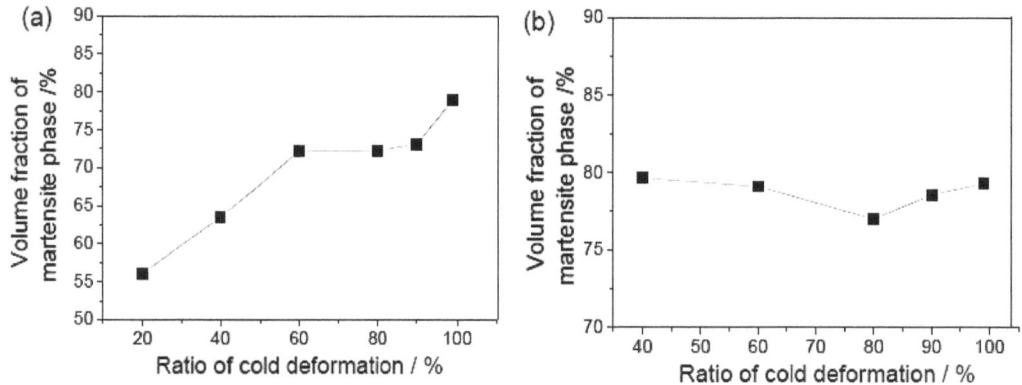

Fig. (13). The relationship between the ratio of deformation and the volume fraction of martensite during cold rolling: (a) direct rolling, (b) cross rolling.

TEM Analysis of Direct-Rolled Structure

Fig. **(14)** displays (a) bright field and (b) dark field micrographs of acicular martensite in the samples direct-deformed by 20%. In the stage of martensite nucleation, these parallel structures with straight boundaries are regularly obtained. Densely distributed dislocations can be observed at the end of this acicular martensite. Since the cold reduction is relatively small, martensite's growth is not very apparent, results in a thin acicular structure, which is about 15nm in thickness shown in Fig. (**14a**). Observed from the dark field image (Fig. **14b**), this parallel distributed martensite is twin martensite. In addition, during the deformation process, deformation twins are found. This deformation twins expands along different directions, whose lath thickness varies between 100nm to 600nm as obviously presented in Fig. (**14c**). Ti-22.5Nb-2Zr-0.7Ta alloy has higher β-phase stability, so its dislocation slipping takes place under relatively large deformation reduction. Fig. (**14d**) shows the TEM structure of the place where dislocation piles up in the specimens rolled by 20% in thickness. Dislocations tangles, tends to form cell structures, as indicated by the white arrows. As the increase of the amount of deformation twins appears during the rolling process, the orientation of grains also tends to change, as already observed in alloys with low reduction (20%).

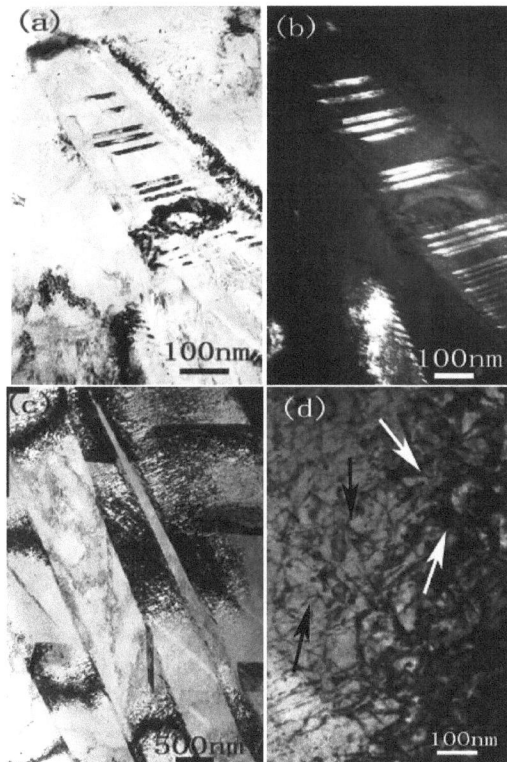

Fig. (14). TEM micrographs of deformed samples deformed by 20%: (a) and (b) the TEM bright and dark field images of stress-induced α'' martensite, respectively; (c) the image of deformation twins; (d) the image of dislocation structure [33].

Fig. (15) presents the TEM microstructure after 60% direct-rolling. In Fig. (15a), along with the accretion of cold deflection, strain-induced martensite distributes densely in the place where dislocation density is relatively high. Compared with the martensite generated under a reduction of 20%, whose average thickness is 15 nm, small acicular martensite with this deformation ratio nucleates continuously to around 50nm. Fig. (15b) shows selected area electron diffraction (SAED) pattern, and diffraction patterns are illustrated in Fig. (15c). In the image, one with open circles is from the [113]β zone axis of the matrix phase, and the solid circles stand for diffraction pattern of strain-induced martensite from the [001]α" zone axis. The illustration reveals that the strain-induced martensite is (110) twin. Diffraction patterns analysis indicates that the orientation relationship between acicular martensite phase and β matrix is identified as [001]α" // [113]β. As for TiNbZrTa alloy, in the process of deformation, dislocations slipping took place under relatively large deformation reduction, and variant martensite collided with each other. These contributed mostly to alloy's deformation. Fig. (15d) shows the TEM structure of the place where dislocation piled up.

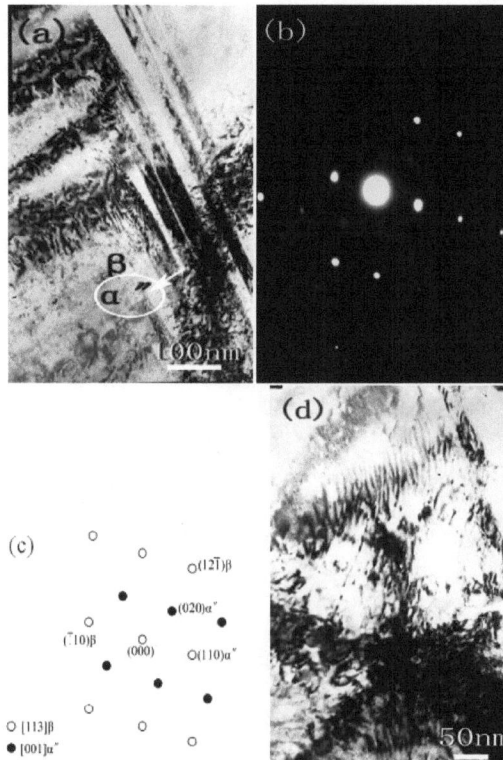

Fig. (15). TEM micrographs of deformed samples deformed by 60%: (a) image of martensite; (b) the corresponding SAED pattern with the reflections from variants of martensite and the β-Ti matrix; (c) the illustration of (b); (d) the image of dislocation structure [33].

In Fig. (**16**), TEM micrographs of deformed samples rolled by 90% are presented. Deformed at greater rate, slipping bands are along particular direction with dense distribution of dislocation cells nearby. Fig. (**16**) indicates the slip process under the effect of shear stress τ. the arrangement features before and after atomic distortion under the function of shear stress are exhibited in Figs. (**16a-b**). It's not difficult to find that in the rolling process atoms move along the direction of shear stress a distance of an integer multiple of lattice spacing, and in this way slipping bands appear.

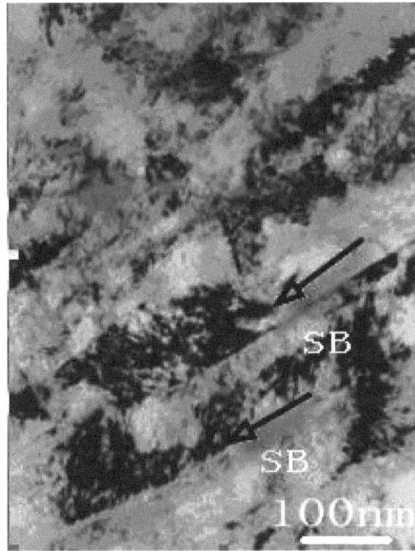

Fig. (16). TEM micrographs of deformed samples deformed by 90%.

In Fig. (**17**), TEM micrographs of deformed samples rolled by 99% are presented. Deformed at greater rate, the material shows characteristic of inter tangled dislocation and cells-like cluster structure. Dislocations' extension and slipping contribute to the great density of dislocation in this condition. Meanwhile, it is clearly seen in Fig. (**17b**) that, shear bands (SB) are along particular direction. During direct-rolling, the shear bands are cut into thinner band by dense dislocation walls. It is also visible in Fig. (**17b**) that the discontinued shear bands' width is about 60nm, which indicates that grains are refined dramatically. And refinement features of grains can be concluded from the ring-shaped diffraction patterns in Fig. (**17c**). Under the interaction between dislocations and shear bands, large plastic deformation occurs continually, meanwhile, achieving grain refinement that the alloy obtains rolled organizations smaller than 100nm.

The cold deformation mechanisms of TiNbZrTa β titanium alloy in direct-rolling process can be inferred from the TEM analysis above. Meanwhile, the stress fields around the dislocation promote the nucleation and growth of strain-induced martensite. When the reduction ratio increases, the martensitic transformation is prevented by dislocations and cell structure, where both the matrix and plate-shaped α" martensite phases are refined, and stress-induced α" martensite phase and deformation twins appear. When the dislocation slipping is restrained in the alloy deformed by 20%, plate-shaped deformation twins contribute much to the plastic deformation. In large reduction dislocation and shear bands slipping contributed most to the plastic deformation.

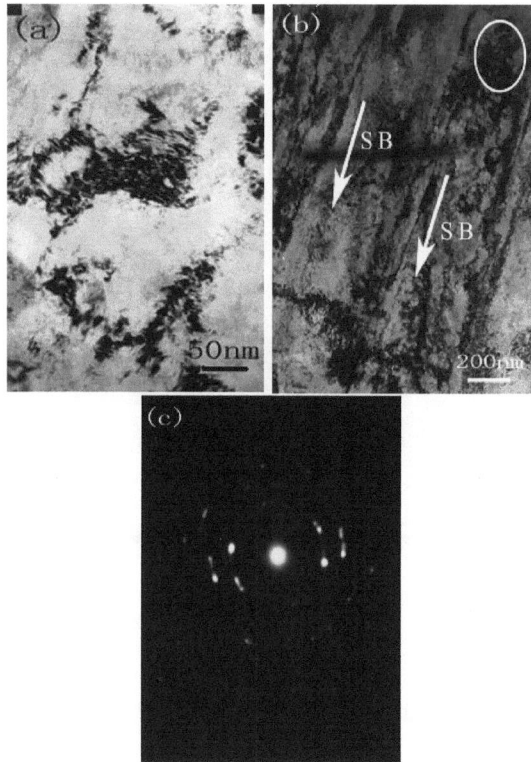

Fig. (17). TEM micrographs of deformed sample rolled by 99%: (a) the TEM bright field image of dislocations; (b) the TEM bright field image of shear bands; (c) the corresponding SAED pattern with the reflections from the β-Ti matrix [33].

Mechanical Properties of Direct-Rolled Specimen

Fig. (**18a**) shows the tensile curves of the specimens after solution and cold deformation at different deformation rates at the room temperature. Seen from the tensile curves of the specimens treated by solution, during tensile processing, when the tensile strength is around 200MPa, obvious tensile platform appears, which is shown in Fig. (**18b**). When the strength reaches to 200MPa, martensite microstructure appears which has a close relationship with the shape memory effect. While when the tensile strength increases to 400MPa, the strength keeps the same value and during this process, the slipping of α" martensite phase contributes much to the plastic deformation. The elongation of the specimen after solution treatment reaches to 23%, which shows excellent plasticity.

In the cold-rolled specimen at the deformation rate of 80%, the tensile strength of the alloy increases to 800MPa. Fig. (**18c**) shows the elongation of the specimen under different deformation rates. The elongation is 15% at the cold deformation rate of 20%. When the cold deformation rate reaches to 40%, the elongation seriously decreases. While when the elongation is between 40% and 80%, the elongation decreases gradually and the good plasticity is obtained which is around 10% at the deformation rate of 80%.

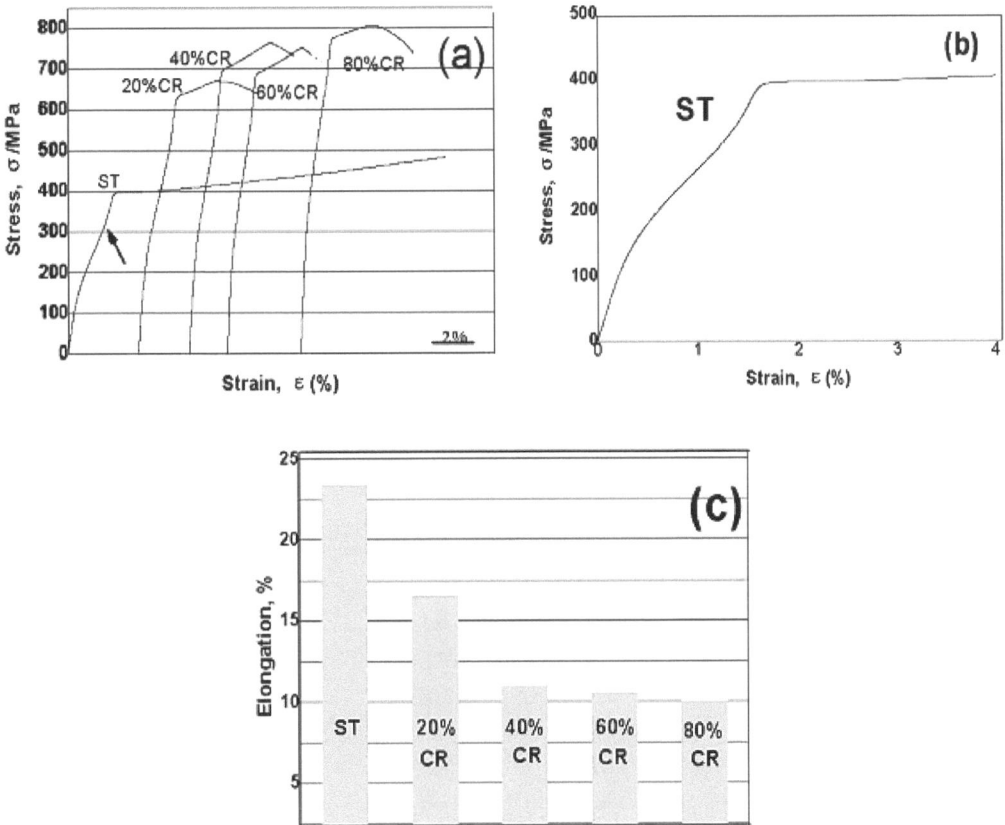

Fig. (**18**). Stress-strain curves (a,b) and elongation to fracture (c) of specimens treated by solution (ST)and cold rolling(CR) [34].

Fig. **(19)** shows the SEM images of the fracture surface of the cold-rolled specimens at the deformation rates of 20% and 80%. Seen from the figure, ductile fracture characteristics appear. Compared with the specimen at 80% deformation rate, because of high elongation, the fracture shows deeper dimples.

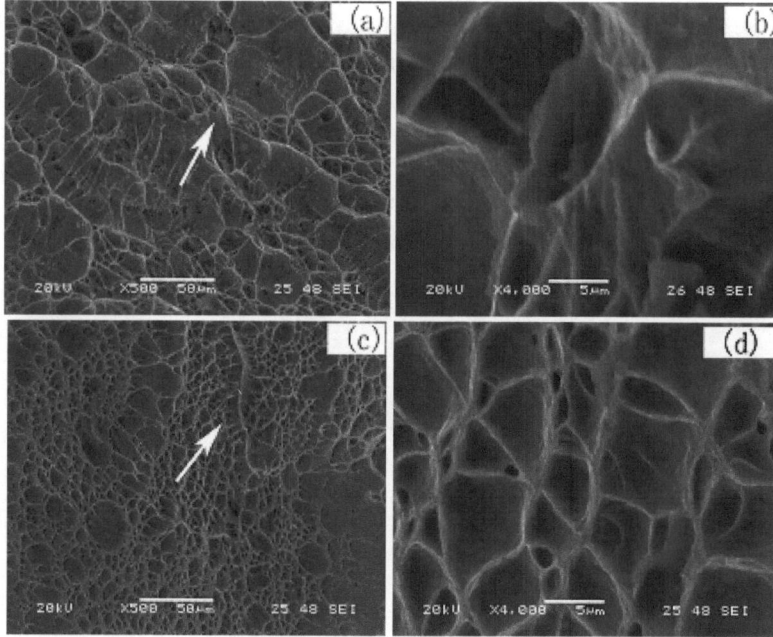

Fig. (19). (a, c) Low-and (b, d) high-magnification SEM images of the fracture surface of 20% (a, b) and 80% (c, d) tensile tested specimens [34].

Mechanical Properties of Cross-Rolled Specimen

Fig. **(20)** shows the stress-strain curve of the cross-rolled specimens at the deformation of 60% along the cross rolling directions (CRD1, CRD2) and 45° from cross rolling direction (45DCR). Seen from the figure, the tensile strength along CRD1, CRD2 and 45DCR is 760.4MPa, 774MPa and 775MPa, respectively. The similar values of the tensile strength along three different directions are obtained. As for the cross-rolled Ti-35Nb-3Zr-2Ta specimen, the anisotropy of tensile strength along different directions of anisotropy is not obvious. In addition, the mechanical properties of the rolled specimens at the deformation rates of 40% and 99% along three directions are measured. Tables **4(a-c)** show the tensile properties of the specimens at the deformation rates of 40%, 60% and 99% along three different directions. It can be seen clearly that the yield strength changes little with the increase of deformation rate and the yield strength is similar along CRD1, CRD2 and 45DCR.

Tensile strength increases with the increase of cold deformation rate gradually, when the cold deformation rate is 99%, the tensile strength increases to 880MPa. Along CRD1, CRD2 and 45DCR, the value of tensile strength is similar. Therefore, the mechanical properties along different directions are similar which are caused by the uniform microstructure along different directions. The same change characteristic of elongation and elastic modulus are obtained which is shown in Table **4**. With the increase of deformation rate, elastic modulus decreases and the value decreases to 50GPa at the deformation rate of 99%.

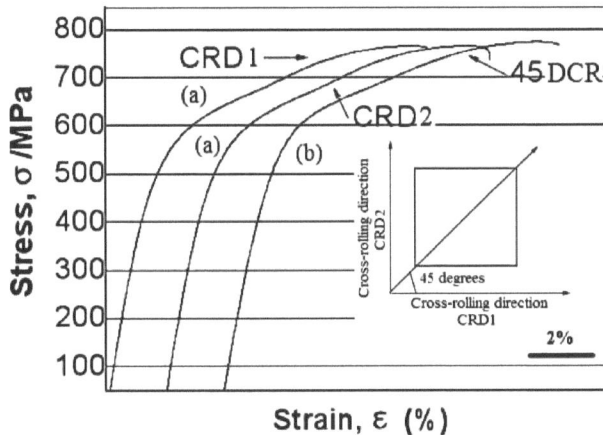

Fig. (20). Stress–strain curves of Ti-35Nb-3Zr-2Ta alloy cross-rolled at a reduction ratio of 60% (a) along with cross-rolling direction (CRD1 and CRD2) and (b) inclined 45 degrees to cross-rolling direction (45DCR) [35].

Table 4. Tensile properties of Ti-35Nb-3Zr-2Ta alloy subjected to a solution treatment and cross-rolling at reduction ratios of 40%, 60%, and 99%:(a) CRD1;(b) 45DCR;(c) CRD2.

(a)				
Sample	**Yield strength (MPa)**	**Ultimate tensile strength (MPa)**	**Elongation (%)**	**Young's modulus (GPa)**
40% reduction	510	749	10	65.2
60% reduction	514	760	9.5	60.0
99% reduction	522	880	9.0	48.8

(b)				
Sample	**Yield strength (MPa)**	**Ultimate tensile strength (MPa)**	**Elongation (%)**	**Young's modulus (GPa)**
40% reduction	513	751	11.3	65.6
60% reduction	519	775	9.9	61.2
99% reduction	529	885	9.5	50.3

(Table 4) contd.....

(c)				
Sample	**Yield strength (MPa)**	**Ultimate tensile strength (MPa)**	**Elongation (%)**	**Young's modulus (GPa)**
40% reduction	515	756	10.2	66.3
60% reduction	516	775	9.4	59.6
99% reduction	523	883	9.0	49.3

SHAPE MEMORY EFFECT OF TI-35NB-3ZR-2TA ALLOY

Fig. (**21**) shows the elastic modulus of the specimens deformed by the reduction of 99% followed by heat treatment at 873K and 1223K. When the heat treatment temperature changes from 873K to 1223K, the elastic modulus increases from 55.1GPa 57.1GPa along the rolling direction. Lower elastic modulus is obtained along the direction which is 45° from the rolling direction. The value is 39.4GPa and 47.4GPa, respectively at these different temperatures.

Compared to the other two directions, the highest elastic modulus is obtained along the direction perpendicular to the rolling direction, which is 70.0GPa and 82.4GPa, respectively. At present, the variation of the elastic modulus of β titanium has been reported by many scholars. For body-centered cubic metals, the values of elastic constants C_{11}, C_{12} and C_{44} determine the elastic modulus. The elastic modulus of E001 along [001] direction can be calculated by the following formula:

$$E001 = (C_{11} + 2\,C_{12})\,(C_{11}-C_{12}) / (C_{11} + C_{12}) \qquad (2)$$

On the formula above, C_{11}-C_{12} is representing the number of vacancy number of valence electrons. The value of C_{11}-C_{12} closes to zero when vacancy number of valence electrons of the alloy ranges from 4.20 to 4.24. And the elastic modulus along to the [001] direction closes to zero. The calculated vacancy number of valence electrons of Ti35Nb3Zr2Ta is 4.257, which means that smallest value of elastic modulus is obtained along the direction of E001.

As for the former research of the elastic modulus of β titanium alloys, the change trend of elastic modulus along different directions is as follows: [001] < [110] <[111] ~ [112] [24, 25]. As shows in Fig. (**21**), as for the specimen treated at 873K, the <001> direction of the grains parallels to the direction which is 45° from rolling direction. Therefore, the minimum value of elastic modulus appears along this direction which is 39.4GPa. Compared to the rolling direction and perpendicular to the rolling direction, the elastic modulus is lower along the rolling direction. The anisotropy of the elastic modulus of the heat treated titanium alloys along the three different directions is caused by the different grain orientations. Along with the direction which is 45° from rolling direction, due to the recrystallized grain orientation of [012] direction, lower elastic modulus (47.4GPa), is obtained. In addition, along the direction perpendicular to the rolling direction, higher elastic modulus is obtained because of the recrystallized grain orientation in [111] direction.

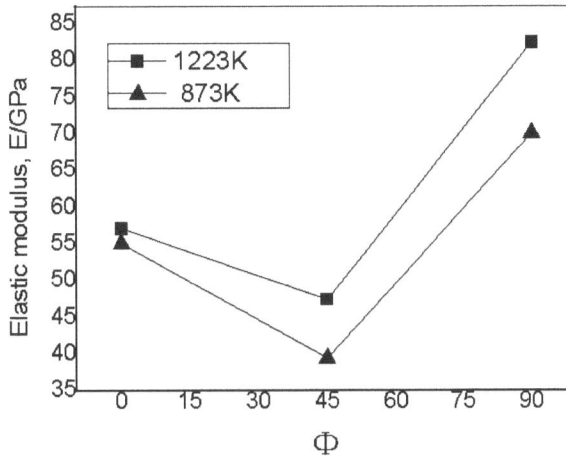

Fig. (21). Elastic modulus along different directions of 99% cold-rolled TiNbZrTa specimens treated at 873K and 1223K [32].

Fig. (**22**) shows the stress-strain tensile curve of the specimen heat treated at 873K. Tensile axis is along the rolling direction, 45° to the rolling direction and perpendicular to the rolling direction, respectively. Tensile strain rate is 2.5%, 4.0%, and 5.5%, respectively. In the first cycle, when the strain rate reaches 2.5%, the loading is unloaded and the operation is repeated at the tensile strain rate of 4.0% and 5.5%, respectively. The superelastic characteristics of the specimens are expressed by εSE and εE. εSE and εE are the strain and pure elastic strain in the process of unloading. The platform stage as indicated by the arrows in the diagram of the tensile yield shows the strain-induced martensitic transformation. It can be seen, the maximum value of the pure elastic strain is obtained along the direction which is 45° to the rolling direction. The value of superelasticity along the rolling direction and perpendicular to the rolling direction is greater than that along the direction which is 45° from the rolling direction. The maximum value of εSE about 1.42% was obtained along the rolling direction at the strain rate of 2.5%. Seen from the stress-strain tensile curve in Fig. (**22**), martensitic transformation yield platform appears along the rolling direction and martensitic transformation can be induced when the tensile strength keeps at a lower value (220MPa). Compared with the direction which is 45° to the rolling direction and perpendicular to the rolling direction, martensite transformation can be induced more easily along rolling direction. Some researchers have reported, when the deformation texture changes from [011] to [001], the martensitic transformation strain decreases gradually. Therefore, from the previous texture analysis, a good superelastic characteristic can be found in the <011> direction of the grain, which is parallel to the rolling direction.

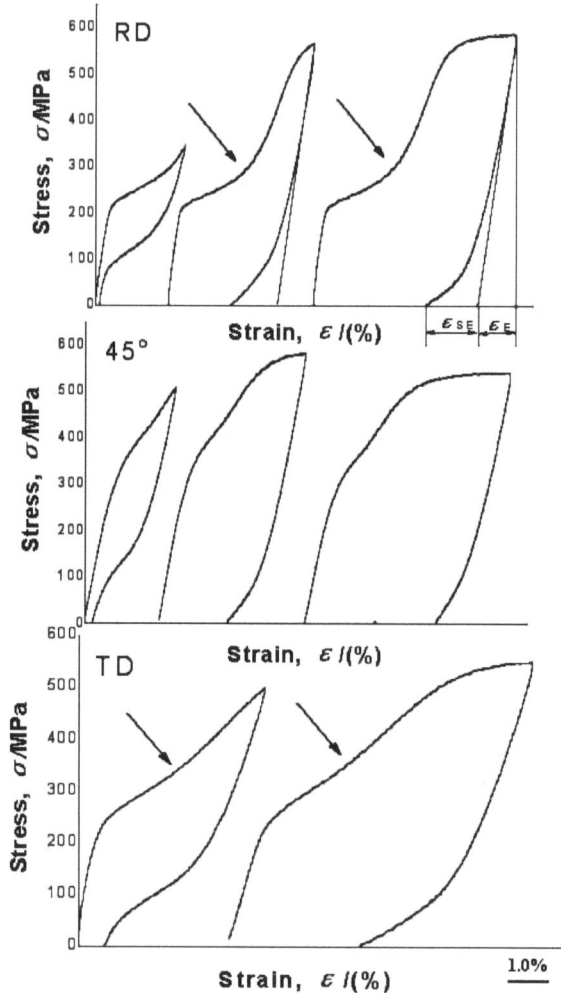

Fig. (22). Stress–strain curves obtained by cyclic loading–unloading tensile tests for the specimen heat treated at 873K for 1.2 ks [32].

Fig. (**23**) shows the stress-strain tensile curve of the specimens treated at 1223K. The place where the arrows marks on the curve shows the shape recovery when the specimen is heated to 523K for 1.2ks after unloading. εr and εTr are the value of total recovery strain and the recovery strain of phase transformation. Fig. (**24**) shows the relationship between εSE, εE, εTr and the tensile directions at the strain ratio of 2.5% and 5.5%. Along the rolling direction, εE shows a higher value, which results from the [011] direction of the recrystallized grain. In addition, because the recrystallized grain parallels to [111], minimum value of the εSE is obtained along the perpendicular to the rolling direction. Similar phenomenon can be observed at both the strain rate of 2.5% and 5.5%. Obvious anisotropy of εTr is found in the specimen with the strain rate of 5.5%. Along the rolling direction, 45° to the rolling direction and perpendicular to the rolling direction, the recovery strain of phase transformation is 2.11%, 2.03% and 1.40%, respectively.

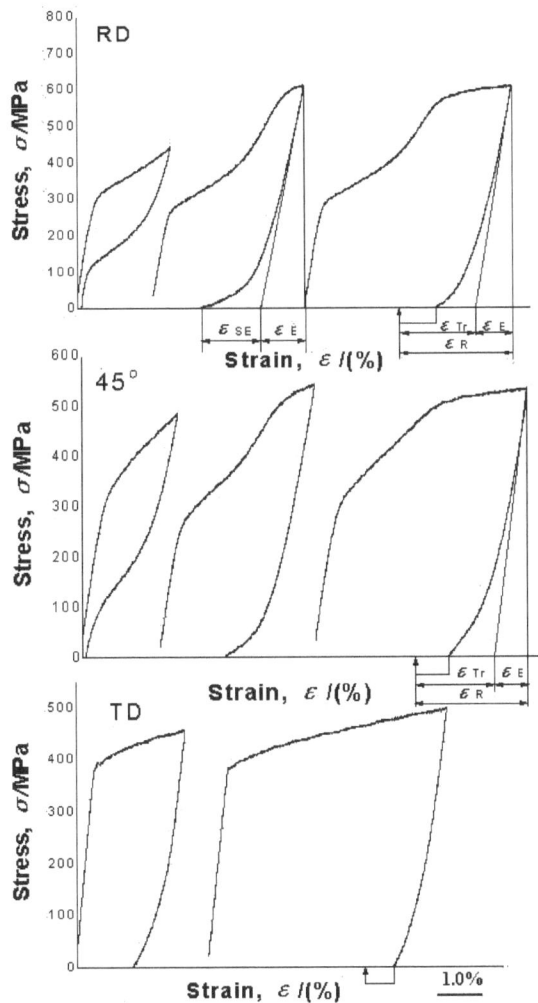

Fig. (23). Stress–strain curves obtained by cyclic loading–unloading tensile tests for the specimens heat treated at 1223K for 1.2 ks [32].

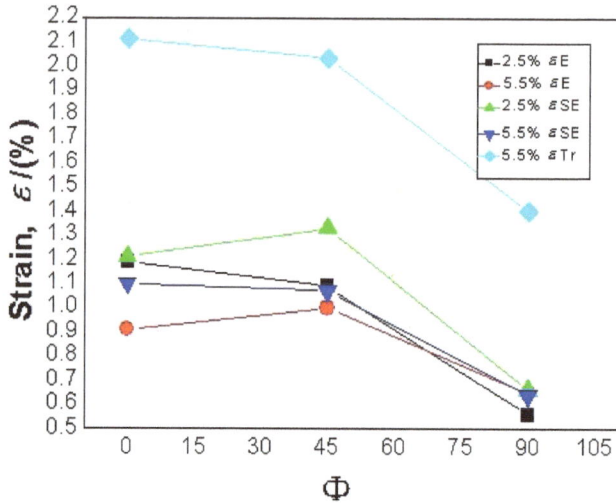

Fig. (24). Φ dependence of εSE, εE and εTr extracted from the stress–strain curves at strain of 2.5% and 5.5% for the specimens heat treated at 1223 K for 1.2 ks [32].

Fig. (**25**) shows the TEM of the unloaded specimen at the strain rate of 5.5%. As shown in the Fig. (**25a**) indicated by the black arrows, strip-shaped deformation twins appear during tensile. In general, in the process of tensile deformation, dislocation slipping occurs which contributes much to the plastic deformation. The appearance of the deformation twins is benefit to the further deformation accompanied with the dislocation slipping.

As shown by the white arrow in Fig. (**25a**), the needle-like martensite twins can be observed clearly in both bright and dark field images. The interface between variants is straight and clear, showing an excellent bonding. Under the condition of low applied stress, this kind of martensite contributes certain plastic deformation.

Fig. (**26**) is the XRD phase analysis of the specimen after tensile experiments. compared with the phase analysis before and after tensile at the strain rate of 5.5%, the diffraction peaks of strain-induced martensite α" phase appears after tensile experiments.

According to the martensitic microstructure analysis in Fig. (**25**), when the stress is applied to the alloy, strain-induced martensite grows up and changes to certain direction, which causes the maximum strain along the direction of the applied stress. When the stress is removed, the recovery of martensite transformation could not appear. When heated to certain temperature, the martensite variants disappear and change to mother phase, which causes shape memory effect.

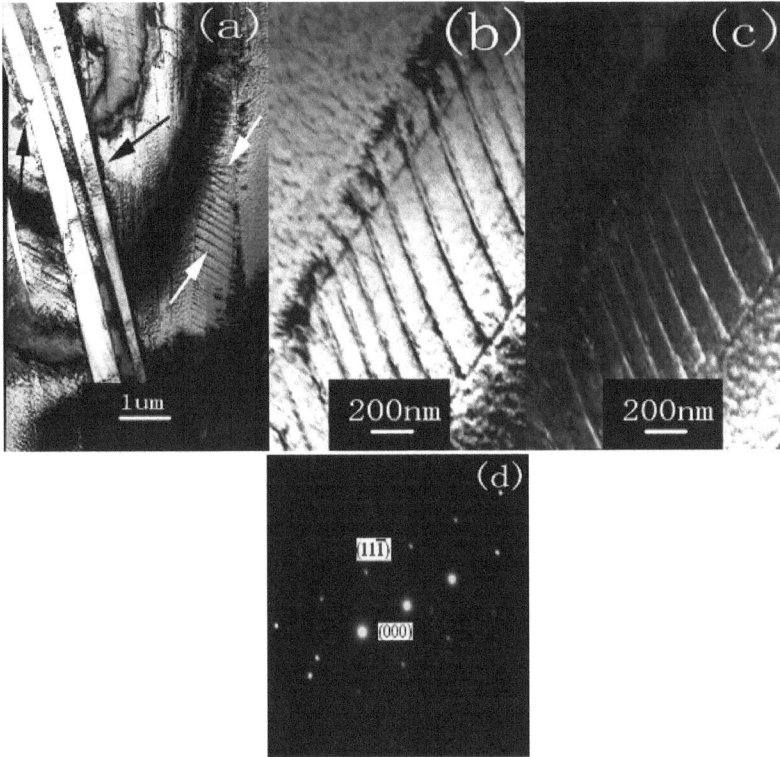

Fig. (25). TEM micrographs of the 5.5% tensile specimen heat treated at 1123K for 1.2ks: (a) image of twin structure; (b) and (c) the TEM bright and dark field images of stress-induced α'' martensite, respectively; (d) the corresponding selected area electron diffraction (SAED) pattern of martensite phases [32].

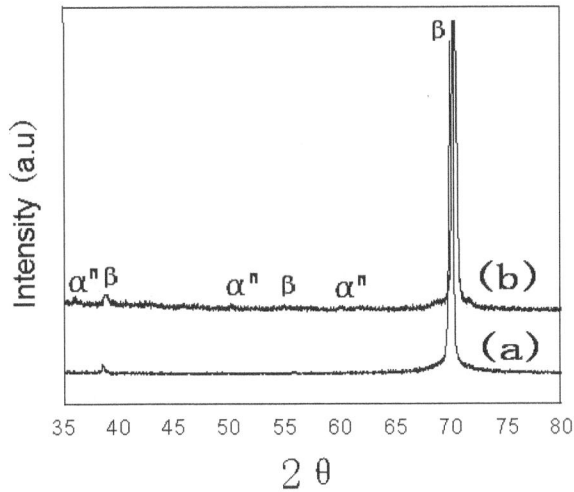

Fig. (26). X-ray diffraction profiles of the specimens: (a) heat-treated specimen at 1223 K for 1.2 ks; (b) 5.5% tensile specimen heat treated at 1123K for 1.2ks [32].

CONCLUSIONS

1. Beta titanium alloys with the elements of Nb,Ta,Zr are being studied as the most important biomedical materials. The TiNbTaZr β titanium alloy with lower elastic modulus and non-toxic alloying elements has much more important application in biomedical field.
2. The Ti35Nb2Ta3Zr β titanium alloy is studied and the alloy has the complex properties of lower elastic modulus, high strength, high elongation and excellent shape memory effect.
3. Compared with direct rolling, cross rolling is benefit to the isotropic of the microstructure and mechanical properties.

CONFLICT OF INTEREST

The author (editor) declares no conflict of interest, financial or otherwise.

ACKNOWLEDGEMENTS

We would like to acknowledge a financial support provided by High Technology Research and Development Program of China under Grant No: 2006AA03Z559, 973 Program under Grant No:2007CB613806, A Foundation for the Author of National Excellent Doctoral Dissertation of PR China under Grant No: 200332.

REFERENCES

[1]   Van NR. Titanium: The implant material of today. J Mater Sci 1987; 22: 3801-11.
       [http://dx.doi.org/10.1007/BF01133326]

[2]   Leventhal GS. Titanium, a metal for surgery. J Bone Joint Surg Am 1951; 33-A(2): 473-4.
       [http://dx.doi.org/10.2106/00004623-195133020-00021] [PMID: 14824196]

[3]   Park JB. Biomater: an introduction 1992.

[4]   Piao M, Miyazaki S, Otsuka K, Norimasa N. Effects of Nb addition on the microstructure of Ti-Ni alloys. Mater Trans, JIM 1992; 33: 337-45.
       [http://dx.doi.org/10.2320/matertrans1989.33.337]

[5]   Yang G. Recent advances in the study of the Ti-Ni-Nb shape memory alloy with wide hysteresis. Rare Met Mater Eng 1998; 27: 322.

[6]   Belyaev SP, Resnina NN, Volkov AE. Influence of irreversible plastic deformation on the martensitic transformation and shape memory effect in TiNi alloy. Mater Sci Eng A 2006; 438: 627.
       [http://dx.doi.org/10.1016/j.msea.2006.02.067]

[7]   Ding XD, Suzuki T, Sun J, Ren X, Otsuka K. Study on elastic constant softening in stress-induced martensitic transformation by molecular dynamics simulation. Mater Sci Eng A 2006; 438: 113.
       [http://dx.doi.org/10.1016/j.msea.2005.12.075]

[8]   Zhang L, Xie CY, Wu JS. Martensitic transformation and shape memory effect of Ti-49 at.% Ni alloys. Mater Sci Eng A 2006; 438: 905.
       [http://dx.doi.org/10.1016/j.msea.2006.02.184]

[9]   Long M, Rack HJ. Titanium alloys in total joint replacement-a materials science perspective. Biomaterials 1998; 19(18): 1621-39.

[http://dx.doi.org/10.1016/S0142-9612(97)00146-4] [PMID: 9839998]

[10]   Rack HJ, Qzai JI. Titanium alloys for biomedical applications. Mater Sci Eng C 2006; 26: 1269-77.
[http://dx.doi.org/10.1016/j.msec.2005.08.032]

[11]   Steinemann SG. Metal implants and surface reactions. Injury 1996; 27 (Suppl. 3): SC16-22.
[http://dx.doi.org/10.1016/0020-1383(96)89027-9] [PMID: 9039349]

[12]   Duerig TW, Albrecht J, Richter D, Fischer P. Formation and reversion of stress induced martensite in
Ti-10V-2Fe-3Al. Acta Metall 1982; 30: 2161-72.
[http://dx.doi.org/10.1016/0001-6160(82)90137-7]

[13]   Ho WF, Ju CP, Lin JH. Structure and properties of cast binary Ti-Mo alloys. Biomaterials 1999;
20(22): 2115-22.
[http://dx.doi.org/10.1016/S0142-9612(99)00114-3] [PMID: 10555079]

[14]   Grosdidier T, Philippe MJ. Deformation induced martensite and superelasticity in a β-metastable
titanium alloy. Mater Sci Eng A 2000; 291: 218-23.
[http://dx.doi.org/10.1016/S0921-5093(00)00921-7]

[15]   Tang X, Ahmed T, Rack HJ. Phase transformations in Ti-Nb-Ta and Ti-Nb-Ta-Zr alloys. J Mater Sci
2000; 35: 1805-11.
[http://dx.doi.org/10.1023/A:1004792922155]

[16]   Banerjee R, Nag S, Stechschulte J, Fraser HL. Strengthening mechanisms in Ti-Nb-Zr-Ta and Ti-M-
-Zr-Fe orthopaedic alloys. Biomaterials 2004; 25(17): 3413-9.
[http://dx.doi.org/10.1016/j.biomaterials.2003.10.041] [PMID: 15020114]

[17]   Kim HS, Kim WY, Lim SH. Microstructure and elastic modulus of Ti–Nb–Si ternary alloys for
biomedical applications. Scr Mater 2006; 54: 887-91.
[http://dx.doi.org/10.1016/j.scriptamat.2005.11.001]

[18]   Zhou YL, Niinomi M, Akahori T. Decomposition of martensite α″ during aging treatments and
resulting mechanical properties of Ti–Ta alloys. Mater Sci Eng A 2004; 384: 92-101.
[http://dx.doi.org/10.1016/j.msea.2004.05.084]

[19]   Zhou YL, Niinomi M, Akahori T. Effects of Ta content on Young's modulus and tensile properties of
binary Ti–Ta alloys for biomedical applications. Mater Sci Eng A 2004; 371: 283-90.
[http://dx.doi.org/10.1016/j.msea.2003.12.011]

[20]   Laheurte P, Eberhardt A, Philippe MJ. Influence of the microstructure on the pseudoelasticity of a
metastable beta titanium alloy. Mater Sci Eng A 2005; 396: 223-30.
[http://dx.doi.org/10.1016/j.msea.2005.01.022]

[21]   Yang G, Zhang T. Phase transformation and mechanical properties of the Ti50Zr30Nb10Ta10 alloy
with low modulus and biocompatible. J Alloys Compd 2005; 392: 291-4.
[http://dx.doi.org/10.1016/j.jallcom.2004.08.099]

[22]   Kim HY, Sasaki T, Okutsu K, *et al.* Texture and shape memory behavior of Ti–22Nb–6Ta alloy. Acta
Mater 2006; 54: 423-33.
[http://dx.doi.org/10.1016/j.actamat.2005.09.014]

[23]   Kim HY. Mechanical properties and shape memory behavior of Ti-Nb alloys. Sendai. JAPON: Jpn
Inst Met 2004; 45: 6.

[24]   Hon YH, Wang JY, Pan YN. Composition/phase structure and properties of titanium-niobium alloys.
Sendai. JAPON: Jpn Inst Met 2003.

[25]   Niinomi M. Fatigue performance and cyto-toxicity of low rigidity titanium alloy, Ti-29Nb-13-
a-4.6Zr. Biomaterials 2003; 24(16): 2673-83.
[http://dx.doi.org/10.1016/S0142-9612(03)00069-3] [PMID: 12711513]

[26]   Saito T, Furuta T, Hwang JH, Chen R. Multifunctional alloys obtained *via* a dislocation-free plastic
deformation mechanism. Washington, DC: ETATS-UNIS 2003.

[27] Mythili R, Saroja S, Vijayalakshmi M. Study of mechanical behavior and deformation mechanism in an α–β Ti–4.4Ta–1.9Nb alloy. Mater Sci Eng A 2007; 454: 43-51.
[http://dx.doi.org/10.1016/j.msea.2006.11.028]

[28] Kuroda D, Niinomi M, Morinaga M, Kato Y, Yashiro T. Design and mechanical properties of new β type titanium alloys for implant materials. Mater Sci Eng A 1998; 243: 244-9.
[http://dx.doi.org/10.1016/S0921-5093(97)00808-3]

[29] Ping DH, Mitarai Y, Yin FX. Microstructure and shape memory behavior of a Ti–30Nb–3Pd alloy. Scr Mater 2005; 52: 1287-91.
[http://dx.doi.org/10.1016/j.scriptamat.2005.02.029]

[30] Ikehata H, Nagasako N, Furuta T, Fukumoto A, Miwa K. First-principles calculations for development of low elastic modulus Ti alloys. Phys Rev B 2004; 70: 174113.
[http://dx.doi.org/10.1103/PhysRevB.70.174113]

[31] Song Y, Xu DS, Yang R, Li D, Wu WT, Guo ZX. Theoretical study of the effects of alloying elements on the strength and modulus of β-type bio-titanium alloys. Mater Sci Eng A 1999; 260: 269-74.
[http://dx.doi.org/10.1016/S0921-5093(98)00886-7]

[32] Wang LQ, Lu WJ, Qin JN, Zhang F, Zhang D. The characterization of shape memory effect for low elastic modulus biomedical β-type titanium alloy. Mater Charact 2010; 61: 535.
[http://dx.doi.org/10.1016/j.matchar.2010.02.009]

[33] Wang LQ, Lu WJ, Qin JN, Zhang F, Zhang D. Microstructure and mechanical properties of cold-rolled TiNbTaZr biomedical [beta] titanium alloy. Mater Sci Eng A 2008; 490: 421-6.
[http://dx.doi.org/10.1016/j.msea.2008.03.003]

[34] Wang LQ, Lu WJ, Qin JN, Zhang F, Zhang D. Influence of cold deformation on martensite transformation and mechanical properties of Ti–Nb–Ta–Zr alloy. J Alloys Compd 2009; 469: 512-8.
[http://dx.doi.org/10.1016/j.jallcom.2008.02.032]

[35] Wang L, Lu W, Qin J, Zhang F, Zhang D. Change in microstructures and mechanical properties of biomedical Ti-Nb-Ta-Zr system alloy through cross-rolling. Mater Trans 2008; 49: 1791-5.
[http://dx.doi.org/10.2320/matertrans.MRA2008040]

CHAPTER 3

Microstructure and Mechanical Properties of Beta Type Ti-Fe Based Alloys

Lai-Chang Zhang[1,*] and **Liqiang Wang**[2]

[1] *School of Engineering, Edith Cowan University, Perth, WA, Australia*

[2] *State Key Laboratory of Metal Matrix Composites, Shanghai Jiao Tong University, Shanghai, P.R. China*

Abstract: Novel non-toxic β-Ti alloys have been developed and used in the next generation of metallic implants to replace the present utilized near α-type CP-Ti and (α+β)-type Ti-6Al-4V alloy in orthopaedic applications. Nevertheless, the vast majority of these newly β-type Ti alloys are containing a substantial concentration of costly elements like Ta, Nb and Zr with high density and high melting points. Therefore, it is highly desirable to design new β-type biomedical Ti alloys composed of non-toxic, low-cost, abundant metals to lower fraction of high-cost elements. Very recently, some serials of Ti-xFe-yTa, Ti-Fe-xNb and Ti-Nb-xFe alloys have been developed by using the DV-Xα molecular orbital method. The mechanical properties of the alloys can be evaluated by studying the effects of Ta, Fe and Nb contents on phase transition, β phase stability and microstructure evolution, and compared with the currently applied biological materials to determine its suitability. In the currently designed alloys, Ti-10Fe-10Ta, Ti-7Fe-11Nb and Ti-11Nb-9Fe display the excellent combination of mechanical properties, which make them more attractive compared with the conventionally used CP-Ti and Ti-6Al- 4V materials for biomedical applications. Compared to CP-Ti and Ti-6Al-4V alloys, a new type of Ti alloy with better performance for orthopaedic applications can be designed by appropriate alloy design.

Keywords: Fracture, Mechanical properties, Microstructure, Nanoindentation, Phase stability, Phase transformation, Shear band, Titanium, Wear.

INTRODUCTION

Due to the rapid increase in aged population and/or traffic accident all over the world, the demand for substituting for dysfunctional load-bearing tissues with artificial parts like hip and knee implants is expanding. A number of metal biomaterials for hard tissue replacement (or orthopaedic implants), such as Co-Cr

* **Corresponding author Lai-Chang Zhang:** School of Engineering, Edith Cowan University, 270 Joondalup Drive, Joondalup, Perth, WA 6027, Australia; Tel: 61 8 63042322; Fax: 61 8 63045811; E-mails: lczhangimr@gmail.com; l.zhang@ecu.edu.au

Liqiang Wang & Lai-Chang Zhang (Eds.)

alloys, titanium alloys and 316L stainless steel, have been developed [1, 2]. However, the metallic biomaterials require the properties of low density, high strength, low elastic modulus, favorable biocompatibility and excellent corrosion resistance. Titanium alloys with these advantages stand out in these biological materials, thereby rendering them ideal hard tissue replacement materials [3 - 16].

Among all the Ti alloys for orthopaedic applications, Ti-6Al-4V is the most widely used for medical implants. However, some studies have pointed out that this type of alloy can release toxic aluminum (Al) and vanadium (V) ions in the human body, which may cause some long-term serious health problems such as Alzheimer's disease [17]. At the same time, its elastic modulus (or Young's modulus) is considerably greater than that of the human bones [18], leading to stress-shielding phenomenon thereby eventually loosening the fixation of implant [3, 19]. Consequently, considerable endeavors have been made in the last decades to develop non-toxic and low-modulus Ti alloys [18 - 21]. Some new β-type Ti alloys containing non-toxic β-stabilizer elements (like Nb, Ta, Zr, Mo, *etc.*) have been developed and prepare for the next generation of biomedical metallic materials [22, 23] in order to replace the widely used commercial alloys like Ti-6Al-4V and commercial purity (CP) Ti in orthopaedic surgery applications [21]. However, the newly-developed β-type Ti alloys, such as Ti-42Nb, Ti-50Ta-20Zr and Ti-24Nb-4Zr-8Sn, contain immensely expensive and heavy elements such as Ta, Nb and Zr, which make these Ti alloys having high density and high melting point, thereby leading to a significant inconvenience in the manufacturing process. This chapter reviews the recent progress in the microstructure and mechanical properties of newly developed β-type Ti-Fe based alloys with the features of non-toxic, low-cost, abundant metals such as Fe, Mn or Sn to minimize the consumption of high-cost elements while exhibiting favorable properties potential for biomedical applications.

TI-FE-TA ALLOYS

Haghighi *et al.* [24] have designed a series of Ti-xFe-yTa (x = 8, 9, 10 wt% and y = 0, 2, 5, 8, 9, 10 wt%) alloys by using $\overline{Bo}$-$\overline{Md}$ map and investigated the effect of Fe and Ta contents on the phase transformation, microstructure and mechanical properties of the designed Ti-Fe-Ta alloys. It can be seen from Fig. (1), that all the seven cast Ti-xFe-yTa alloy samples with different Fe and Ta contents have only the body-centrer cubic (BCC) β phase and the orthogonal α" phase. By analysis of each peak from the XRD patterns can calculate the volume fraction (V_f) of the β and α" phases in the studied alloys. For example, the β phase of Ti-8Fe in the XRD spectrum is accounted for 76% and 24% for α" phase. It was also found that the volume fraction of α" phase slightly decreases after adding 5 wt% (weight percent) and 8 wt% of Ta in Ti-8Fe alloy, and the volume fraction of β phase increases to 81% and 85%, respectively. In the case of increasing 1% Fe based on Ti-8Fe alloys (*i.e.*, Ti-9Fe-xTa, x = 2, 5 and 9), the XRD patterns show a higher peak of β phase, causing a relatively higher volume fraction of the β phase compare to the Ti-8Fe alloys. Subsequently, with the further increase in Fe and Ta contents, the volume fraction of α" phase is gradually decreasing. Finally, the volume fraction of α" phase in Ti-10Fe-10Ta alloy reaches the lowest value (V_f = 3%) in all examined samples. The reason for this phenomenon is that with increase in the content of β-stabilizer elements Ta and Fe in the alloy, the β→α" martensitic transformation tendency is reduced and therefore the martensitic start temperature [25, 26], M_s, hence reduces the martensite phase in the Ti-Fe-Ta alloys after cooling to room temperature. Similar results were also observed in the studies of other β stabilizing Ti alloy such as Ti-22Nb-xTa [27], Ti-5Cr-xFe [17] and Ti-10Mo-xSi [28].

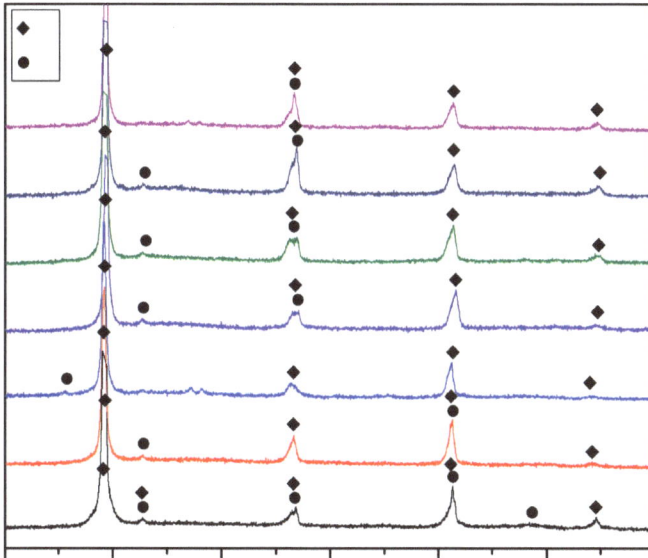

Fig. (1). XRD patterns of as-cast Ti-xFe-yTa alloys (x = 8, 9, 10 and y = 0, 2, 5, 8, 9, 10) [24].

Fig. (**2**) shows the optical microstructure of seven different compositions for Ti-xFe-yTa alloy. As shown in Fig. (**2a**), the Ti-8Fe alloy exhibits a dual phase microstructure which contains the β phase and the precipitated α" phase in β matrix. When the β-stabilizing element Ta is introduced into the Ti-8Fe alloy by 5wt% and 8wt%, respectively, as shown in Fig. (**2b**) and (**c**). The amount of α" phase is significantly reduced, and the α" phase in Fig. (**2c**) is less than the alloy containing 5wt% β-stable element Ta, particularly. Similarly, Fig. (**2d**) indicates that the dual phase is formed in Ti-9Fe-2Ta alloy when additional 1wt% β-stabilizing element Fe and 2wt% Ta are added to the Ti-8Fe alloy. It can be seen from the figure that the amount of α" phases is also relatively large, and at the same time, the β grains are becoming more pronounced. Further addition of Ta enhances the β phase stability as shown in Fig. (**2e**) and (**f**), while also reduces the volume fraction of α" phase. The microstructure of Ti-10Fe-10Ta alloy in Fig. (**2g**) basically do not show any trace of α" phase, although a small amount of α" phase is present in the XRD pattern. As such, the entire alloy is dominant by the β phase structure. The above results show that the stability of the β phase is enhanced with the increase in Fe and Ta content, while the possibility of forming α" martensite is reduced after quenching. This result is also mentioned in other reports [29, 30].

It has been reported that the phase and the corresponding volume fraction present in the microstructure of the Ti alloy affect its mechanical properties [31, 32]. Haghighi *et al.* [24] have also studied the effect of the contents of various elements on the mechanical properties, including the compressive strength, plastic strain, elastic modulus and hardness, of the studied Ti-xFe-yTa alloys. The Fig. (**3a**) summarizes the histogram of Vickers hardness and elastic modulus of seven different samples with β-phase stabilizer elements. It can be seen from the figure that the average Vickers hardness and elastic modulus of the Ti-xFe-yTa alloys are significantly reduced with increasing the weight percent of β-phase stabilizing elements Fe and Ta. According to the previous microstructural characteristics, this phenomenon may be related to the reduction of α" martensite phase present in the microstructure. For the Ti-8Fe alloy, the volume fraction of the α" phase is 24%, its hardness (467Hv) and elastic modulus (128GPa) are the highest among all the studied Ti-xFe-yTa alloys, which are justified by the highest content of α" phase in the tested sample. On the contrary, the hardness and elastic modulus values of the Ti-10Fe-10Ta alloy with the dominant β-phase microstructure reach the minimum value of 343 Hv and 92 GPa, respectively, in all the tested samples. For the Ti-9Fe-2Ta alloy with a high volume fraction of α" martensite phase (21%), it is not difficult to observe an increasing trend in Vickers hardness and elastic modulus values, which may be attributed to the alloying effect of Fe and a small amount of Ta (2 wt%). Meanwhile, the measured Vickers hardness has close results compared to other biomedical Ti alloys. For example, Ti-10Mo-xNb (x = 3, 7, 10) alloy with a Vickers hardness values in a range of 394-441 Hv [33], and 360-435 Hv [34] for Ti-7.5Mo-xFe (x = 0.1-7) alloys. In addition, in the studied Ti-Fe-Ta alloys [24], all alloys have higher value of Vickers hardness, and smaller Young's modulus value than the commercially used biomedical Ti alloy like Ti-9Fe-5Ta, Ti-9Fe-9Ta and Ti-10Fe-10Ta alloys, *i.e.* CP-Ti (190 Hv, 119 GPa) [34, 35] and Ti-6Al-4V (294 Hv, 110-140 GPa) [36, 37].

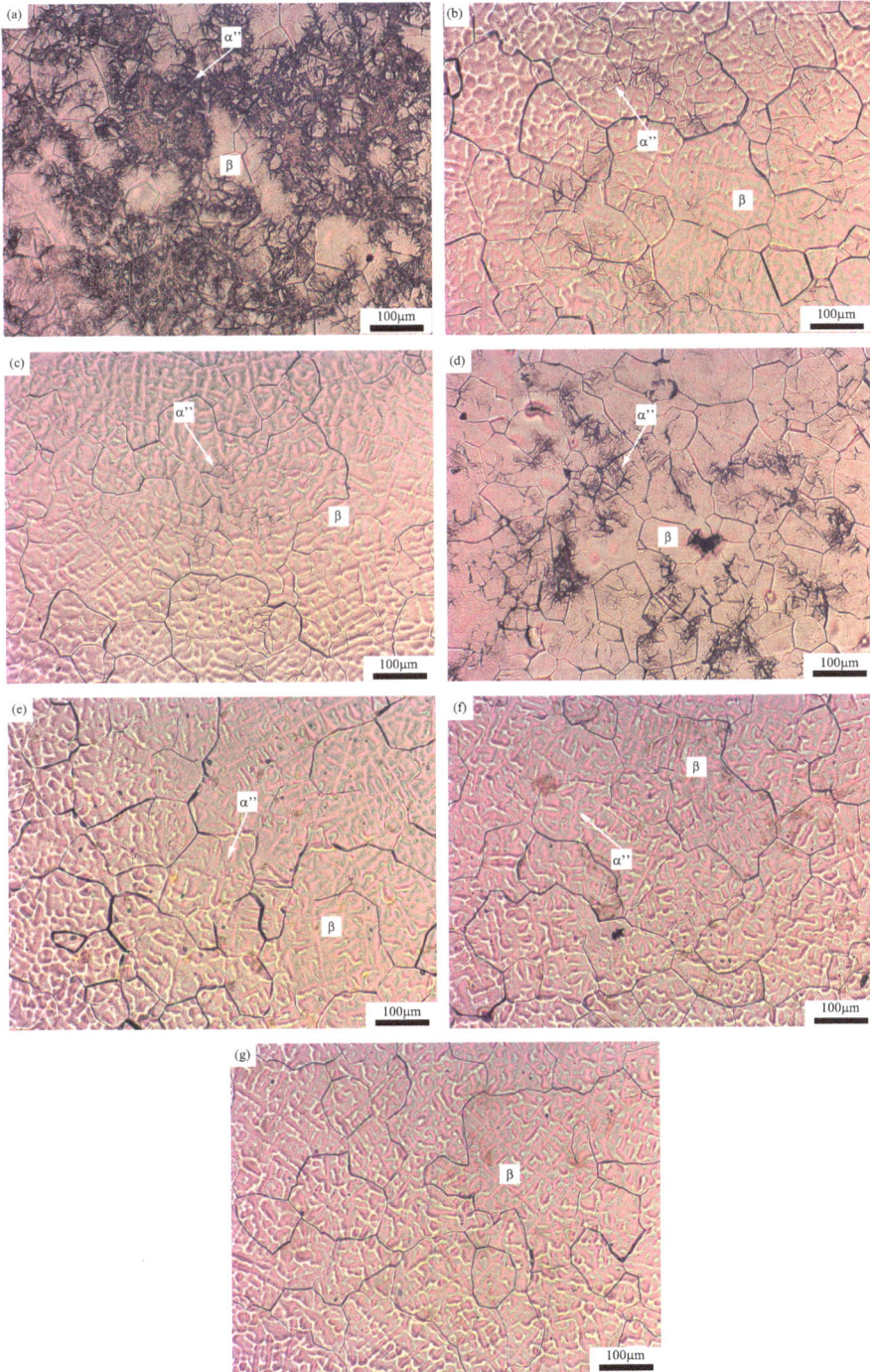

Fig. (2). Optical micrographs of Ti-xFe-yTa alloys: (**a**) Ti-8Fe, (**b**) Ti-8Fe-5Ta, (**c**) Ti-8Fe-8Ta, (**d**) Ti-9Fe-2Ta, (**e**) Ti-9Fe-5Ta, (**f**) Ti-9Fe-9Ta and (**g**) Ti-10Fe-10Ta [24].

Fig. (**3b**) shows the results of the compression testing, which indicate that a relatively high ductility in compression in all test alloys except for Ti-8Fe, and compared with the original shape, the deformation is risen to around 40% without any cracks and fracture. The illustration in Fig. (**3b**) shows the comparison of the Ti-10Fe-10Ta alloy before and after compression, which also confirms that there is no crack at the circumference of the alloy. In addition, these large strains indicate excellent machinability at room temperature [38].

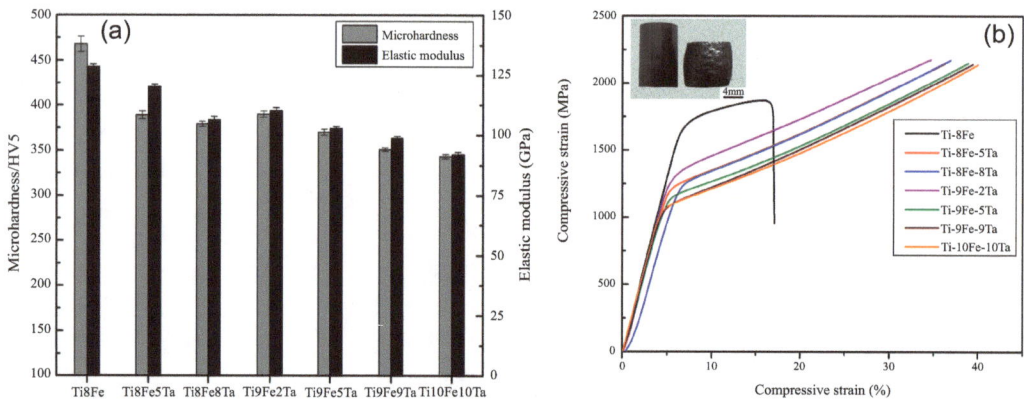

Fig. (3). Mechanical properties of Ti-xFe-yTa alloys: (**a**) Vickers hardness and elastic modulus, (**b**) Compressive stress-strain curves (the inset is the sample before and after the compression test) [24].

TI-FE-NB ALLOYS

In the development of the high strength titanium alloys, Fe is a non-toxic, resource-rich and low-cost β-stablizer element [39 - 41], while in the implants application, the addition of another β-stabilizing element Nb can significantly reduce the Young's modulus of titanium alloy [42, 43]. Nevertheless, relatively limited literature exists on the properties of Ti-Fe alloys with addition of Nb element. Ehtemam-Haghighi *et al.* [44 - 46] have studied the influence of Nb content on the phase transition, specifically on the stability of β phase and mechanical properties along with the wear response of a new series of Ti-7Fe-xNb (x = 0, 1, 4, 6, 9 and 11 wt%) alloys. These alloys are studied to assess its applicability for implant applications.

It is clear from Table **1** that with the Nb content increasing in the Ti-7Fe-xNb alloy with small increments, the volume fraction of the α" phase has a reduction trend from the initial maximum amount of 38%. In this process, when the Nb content is increased from 1wt% to 4wt% and 6wt%, separately. The α" phase is prominently inhibited. When the Nb content is rising to 9wt%, the precipitation of α" phase is almost restricted. Finally, the α" phase completely disappears therefore the retention of full β phase is obtained when adding 11wt% of Nb to the Ti-7Fe alloy.

Table 1. The phase distribution and β stability indicators of as-cast Ti-7Fe-xNb alloys [46].

Alloy composition (wt%)	α″ phase V_f(%)	β phase V_f(%)	β stability indicators		
			$\overline{Bo}$	$\overline{Md}$	Mo_{eq}
Ti-7Fe	38	62	2.781	2.357	17.500
Ti-7Fe-1Nb	34	66	2.783	2.356	17.777
Ti-7Fe-4Nb	18	82	2.787	2.355	18.611
Ti-7Fe-6Nb	10	90	2.791	2.353	19.166
Ti-7Fe-9Nb	4	96	2.796	2.352	20.000
Ti-7Fe-11Nb	-	100	2.800	2.350	20.556

It is not difficult to find that the β-phase stability has a relation with the α-phase transformation in the study of Ti-7Fe-xNb alloy. Based on the DV-Xα cluster method [47] and molybdenum equivalency (Mo_{eq}) parameter [48], the stability of β phase can be easily predicted. In general, decreasing $\overline{Md}$ (metal d-orbital energy level) and increasing $\overline{Bo}$ (the bond order), Mo_{eq} [48, 49] can increase the stability of β phase. So one can do another analysis with regard to the β-phase stability for Ti-7Fe-xNb alloy. The values of Mo_{eq} are higher than the range of getting β phase stability, *i.e.* 10 wt%. Hence, all of the alloys can be classified into the metastable β category [36]. It should be noted that the martensitic phase can also be formed during the rapid cooling of the metastable β alloy [50]. As shown in Table **1**, the Ti-7Fe alloy containing a large amount of α″ phase exhibits the lowest Mo_{eq} value in all of the Ti-7Fe-xNb alloys studied. With increasing Nb content, the values of Mo_{eq} and $\overline{Bo}$ increase, but $\overline{Md}$ decreases, which verifies that the transformation of α″ martensite from β-phase is inhibited with the improvement of the β-phase stability in metastable alloys [51]. Furthermore, the stability of β phase have reached the maximum in Ti-7Fe-11Nb alloy without formation of any α″ martensite. This is supported by the volume fraction values of the present phases for the Ti-Fe-Nb alloys as shown in Table **1**.

Fig. (**4**) shows the yield strength and plastic strain curve of the different concentration of Nb in Ti-7Fe-xNb alloys. It can be seen from the figure that the yield strength and plastic strain of Ti-7Fe-xNb alloy have a specific trends with increasing the percentage of Nb element. Also, the compressive yield strength of the alloy with Nb content of 11% is the lowest, but the test result is also higher than that of CP-Ti (552MPa) [52] and Ti-6Al-4V (970MPa) [53]. Additionally, the Nb content has successfully affected the compressive yield strength and plastic strain of those alloys. However, the Ti-7Fe alloy has exhibited the highest yield strength of 1847 MPa and lowest plastic strain of 2% in this case. Unfortunately this alloy is the most unstable alloy with Mo_{eq} value of 17.5 wt%. This can be attributed to the presence of highest volume fraction of α'' martensite in its microstructure (Table **1**). Since the BCC crystal structure of β phase contains a relatively larger number of slip systems than the orthorhombic crystal structure of α'' phase, the plastic deformation of the α'' phase requires a higher stress than that for the β phase matrix [54]. Conversely, as shown in Fig. (**4**), with the increase in Nb content, Mo_{eq} is enhanced, which would lead to a gradual decrease in the content of high-strength brittle α'' phase and increasing the β phase stability, thereby resulting in significantly improved plastic strain (Table **1**). Therefore, although the highest content of Nb is in the β-phase stability Ti-7Fe-11Nb alloy, its yield strength is even lowest. Nevertheless, the yield strength is still higher than that of other Ti-based biomaterials like CP-Ti and Ti-6Al-4V as discussed above. As this higher yield strength increases the ability of the alloy to change its permanent deformation, this brings the gospel to the majority of patients [55]. In addition, the Ti-7Fe-11Nb alloy exhibits the highest plastic strain (38%) under the maximum loading capacity of 100 kN during compression test due to the high strength of the alloy and large plasticity of β phase structure [56].

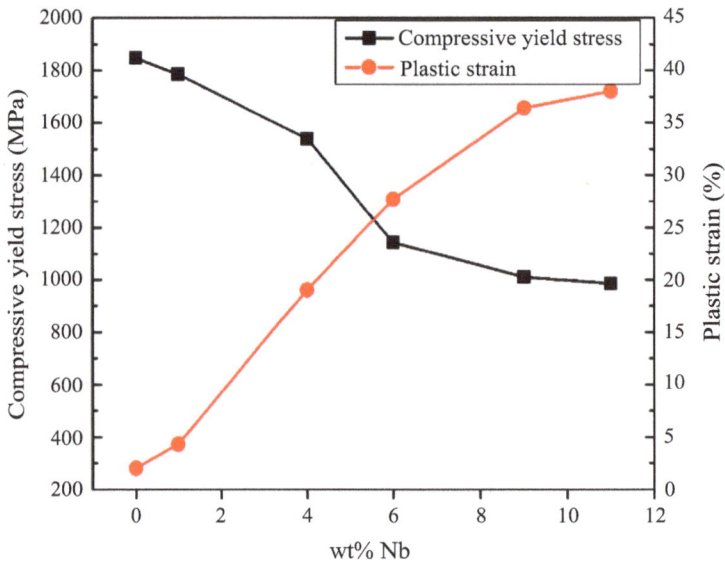

Fig. (4). Compressive yield stress and plastic strain of Ti-7Fe-xNb alloys [46].

The Young's modulus of the studied Ti-7Fe-xNb alloys decreases with the increasing the Nb content as depicted in Fig. (**5a**). Among all studied Ti-7Fe-xNb alloys, the Ti-7Fe alloy presents the largest Young's modulus (129 GPa) because it contains the least stable β phase in microstructure with the lowest Mo_{eq} value of 17.5 wt% and the highest proportion of α″ phase. When separately adding 1 and 4 wt% Nb to Ti-7Fe-xNb alloys, the values of Young's modulus show a slight decrease. Especially when the Nb content is increased to 6wt%, the Young's modulus is greatly reduced and subsequently increases the stability of the β phase. The cause of this phenomenon may be associated with the possible formation of α″ phase after quenching [24]. As shown in Table **1**, except the full β phase of Ti-7Fe-11Nb alloy with highest Nb content, the α″ phase is present in all of the studied alloys. Particularly in the alloys Ti-7Fe, Ti-7Fe-1Nb and Ti- 7Fe-4Nb with a lower Nb content, the proportion of α″ phase ranges between 18% and 38%. One can conclude that with the increase in the weight percent of β-stabilizing element Nb, the corresponding Young's modulus decreases gradually, and reaches the minimum limit of 84 GPa in Ti-7Fe-11Nb alloy with a Mo_{eq} value of 20.56 wt%. Moreover, the obtained results of Young's modulus demonstrates the relationship between various phases of Ti alloys, where the Young's modulus decreases with the phases in the order of β > α″ > α′ > ω [57, 58].

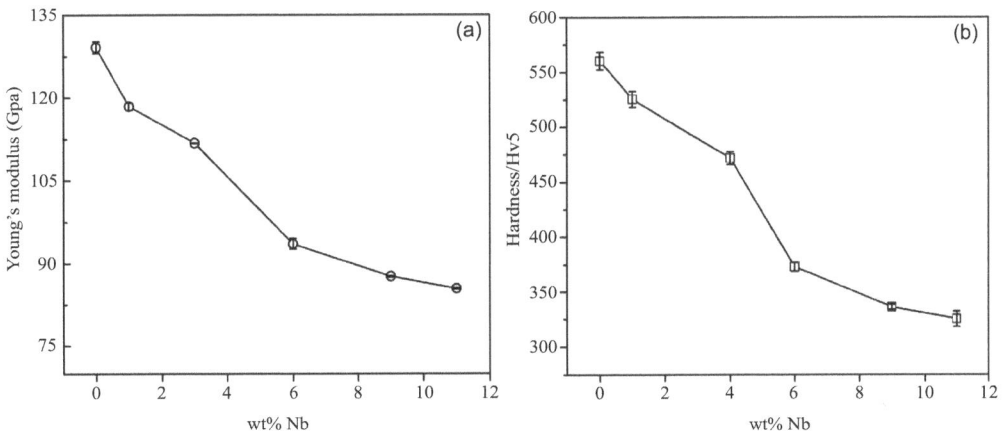

Fig. (5). (**a**) Young's modulus and (**b**) Hardness of Ti-7Fe-xNb alloys [46].

As displayed in Fig. (**5b**), the Vickers hardness has a similar tendency as the Young's modulus does. Similarly, the contents of Nb can also influence the Vickers hardness of the studied alloy. Apparently, at the highest Nb content, the lowest hardness is found to be 325HV, even higher than those for the commercially used biological materialism such as CP-Ti and Ti-6Al-4V with the Vickers hardness of 190Hv [34] and 294Hv [37], respectively. The highest Vickers hardness of 560 Hv can be observed in the Ti-7Fe alloy with the largest α″ content of 38%. With the increase in the β-stablizing element Nb, the Vickers hardness also decreases accordingly. Also, the Vickers hardness drastically reduces to 375 Hv at 6wt% of Nb content. Finally, the Vickers hardness drops to a minimum value of 325 Hv when the β-phase content reaches the peak. This again demonstrates that α" martensitic is more conducive to improving the hardness of alloy than the β phase [57]. Therefore, the reason why the Vickers hardness decreases with the increase in Nb content in Ti-7Fe-xNb alloy can also be explained by the decrease of α" phase content as presented in Table **1**.

Ehtemam-Haghighi *et al.* [59] have also investigated the influence of the Nb concentration on the wear rate variations of the Ti-7Fe-xNb alloys. The declining trend of hardness shown in Fig. (**5b**) upon Nb addition indicates that the wear resistance of the Ti-7Fe-xNb alloys reduces, which demonstrates the increase in wear rate of the alloys as shown in Fig. (**6**). As the Ti-7Fe alloy possesses the highest amount of α" phase, hence it presents the highest hardness and the lowest wear rate (3×10^{-15} m^3/m). Upon addition of Nb, the wear rate increases from 1×10^{-14} m^3/m for Ti-7Fe-1Nb to 1×10^{-13} m^3/m for Ti-7Fe-11Nb alloy with a single β phase microstructure. It is worth noting that the Ti-7Fe-xNb alloys possess a desired hardness than the extensively used Ti alloys like CP-Ti (190 Hv) [34] and Ti-6Al-4V (294 Hv) [37]. Therefore, it is expected that the studied alloys [59] will present a better wear resistance than the commercially used Ti alloys. For example, the wear rate of cast CP-Ti is 1.55×10^{-12} m^3/m under a similar test condition [60]. Overall, an alloy with higher wear resistance represents a lower wear debris production, thereby reducing the possibility of implant loosening while increasing its service lifetime [61].

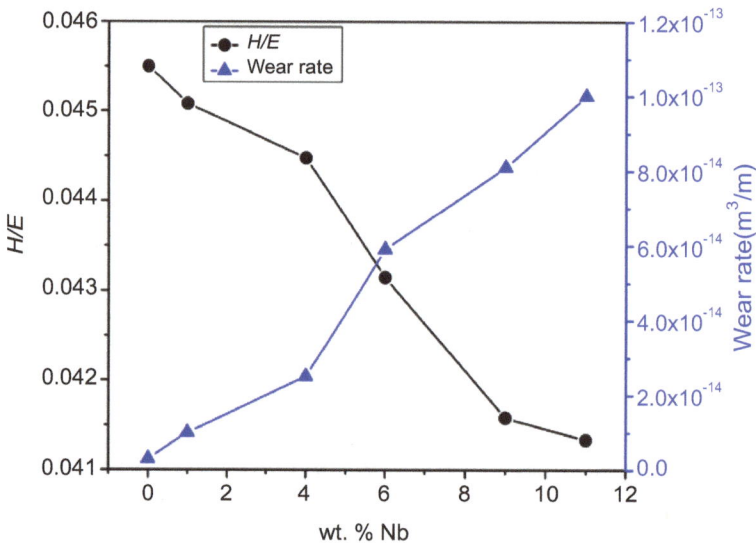

Fig. (6). Wear rate of the Ti-7Fe-xNb alloys [59].

Based on DV-Xα molecular orbital method, Ehtemam-Haghighi *et al.* [44, 45] have also investigated the mechanical properties of a newly designed Ti-11Nb-xFe (x = 0.5, 3.5, 6, 9 wt%) alloys. The mechanical properties of the new series Ti-11Nb-xFe alloys are summarized in Table **2**. It can be inferred from Table **2** that the Fe content has a certain influence on the compressive yield strength and the plastic strain of the studied alloys. The minimum content of the Fe (with the minimum β phase stability indicator Mo_{eq}=4.31) in Ti-11Nb-0.5Fe alloy projects the lowest combination of yield strength and plastic strain with values of 796 MPa and 18%, respectively. This can be ascribed to the large amount of hexagonal close packed (hcp) α phase ($V_{f,α}$ = 83%) present in its microstructure. Continually, the formation of α phase is further restrained with increase the β-stabilizer Fe content from 0.5wt% by 3wt%, which enhances both the yield strength and plastic strain of the alloy. All the changes of yield strength and plastic strain can be attributed to the limited deformation capability of hcp crystal structure, so the β phase exhibits a more ductile behavior than α phase [62, 63]. This behavior can be ascribed to the solid solution strengthening caused by a higher solubility β-stabilizing elements like Nb and Fe in β phase [64]. So, the alloys containing more β phase display a higher strength than the α phase based alloys [65]. As such, the β-phase alloy has better strength and ductility than the alloy containing more α phase [56]. When the addition of 3% Fe in the initial α + β type Ti-11Nb-0.5Fe alloy, the volume fraction of β phase increases, therefore the yield strength and plastic deformation of the alloy are improved. As seen from Table **2**, with increasing the content of Fe to 6 and 9 wt% of Ti-11Nb-xFe, the volume fraction of β phase in the alloy is gradually increased to 90% and 100%, respectively. Also, the plastic deformation reaches a maximum value of 38% in the Ti-11Nb-9Fe alloy. This represents that the alloy will be well machinable at room temperature [24]. In addition, the yield strength of Ti-11Nb-xFe alloy is further enhanced with the increase in Fe content, which indicates that the ability of the alloy to resist permanent deformation is further strengthened. As mentioned above, this can be beneficial to the patient when this material is used as the implant material [55]. However, the yield strength of single β phase structure Ti-11Nb-9Fe alloy (1078 MPa) is slightly lower than the α + β dual-phase alloy of Ti-11Nb-6Fe (1137 MPa), which contains 10% α" phase in volume fraction, and the result is even higher than that of the commercially used Ti-based biomaterials such as CP-Ti (552 MPa) and Ti-6Al-4V (970 MPa) [24].

One can find that the Ti-11Nb-0.5Fe alloy has contained the highest concentration of α phase as one of the four studied samples, which shows the minimum β phase stability (as shown in Table **2**). The Young's modulus has a decreasing trend with gradual increase in the Fe content, but its β-phase stability and Vickers hardness are progressively increased at the same time. Similar to the Young's modulus results, the Vickers hardness of the Ti-11Nb-xFe alloy reaches a maximum value of 357Hv at the Fe concentration of 6wt%, and it slightly decreases as the Fe concentration increases further to 9wt% (334 Hv). But this result is still greater than the ones for the commercially used biomaterial CP-Ti (190Hv) and Ti-6Al-4V (394Hv), even smaller Young's modulus (104 GPa and 114 GPa for CP-Ti and Ti-6Al-4V, respectively) [24], except for the Ti-11Nb-0.5Fe alloy due to its least β phase stability. In general, the β-phase alloy usually has a higher Vickers hardness and a smaller elasticity compared to the alloy with α and (α+β) microstructure [63, 66, 67]. In other words, by adding a β-phase stabilizing element and controlling its concentration, the microstructure of the alloy will change from α/(α + β) phase to β phase thereby enhancing the alloy's Vickers hardness and reducing its elastic modulus, respectively. This also demonstrates the test results of Ti-11Nb-xFe (x = 3.5, 9) alloy. This verdict is also verified by several similar alloys mentioned in the reports such as Ti-5Nb-xFe (x = 1-5 wt%) [1], Ti-10Zr-X (X = Nb, Mo, Cr and Fe) [68] and Ti-22Nb-X (X = Fe, Sn, Ta, Zr, Mo and Si) [69]. The above results indicate that the Young's modulus and Vickers hardness can be successfully tailored by adding the suitable β-stabilizing element Fe to the Ti-11Nb-xFe alloy.

Table 2. The summarization of mechanical properties for the new series Ti-11Nb-xFe alloys. σ_y: yield strength; ε_p: plastic strain; E: elastic modulus; and H: hardness [45].

Alloy nominal composition (wt%)	σ_y (MPa)	ε_p (%)	E (GPa)	H (Hv5)
Ti-11Nb-0.5Fe	796	18	109	278
Ti-11Nb-3.5Fe	932	29	101	303
Ti-11Nb-6Fe	1137	36	89	357
Ti-11Nb-9Fe	1078	38	82	334

Furthermore, the elastic energy is another indicator of the relative bearing capacity of the alloy. As shown in Fig. (**7a**), it is usually the area of the triangular formed by yield stress, elastic strain and stress-strain curve in the elastic deformation stage. The mathematical expression is $\delta_e = \varepsilon_e \cdot \sigma_y/2 = \sigma_y^2 /(2E)$ [70], where δ_e is the elastic energy, ε_e is the elastic strain, σ_y and E are yield strength and Young's modulus, respectively. Based on this calculation method, the elastic energy of the series of Ti-11Nb-xFe alloys can be calculated and compared with some other commercially used Ti-based alloy, as shown Fig. (**7b**) [19, 67]. It is easy to see that the elastic energy of Ti-11Nb-xFe alloy is affected by Fe content. Ti-11Nb-6Fe and Ti-11Nb-9Fe are located at the highest level in the graph due to the higher yield strength and the lower modulus of elasticity measured in Table **2**. However, since the Young's modulus of Ti-11Nb-9Fe alloy is lower than that of Ti-11Nb-6Fe as discussed above, the elastic energy of Ti-11Nb-9Fe alloy displayed in Fig. (**7b**) is also slightly lower than that of Ti-11Nb-6Fe alloy. According to Fig. (**7b**), among all the pointed Ti-based alloys, the elastic energy of Ti-11Nb-6Fe and Ti-11Nb-9Fe are prominently higher, which indicates that the higher β-phase concentration of Ti-Nb-Fe alloy could withstand a higher load than other Ti alloys.

Based on the above achieved results, it can be found that the volume fraction of β phase in Fe-TiNi-xFe alloy has a significant effect on the mechanical properties such as hardness, Young's modulus, strength, plasticity and elastic properties. The volume fraction of β phase is also determined by the content of Fe. Furthermore, the Ti-11Nb-9Fe alloy with a single β-phase microstructure shows a combination of excellent mechanical properties as compared to other Ti-based biomaterials. For example, the Ti-11Nb-9Fe alloy exhibits the lowest Young's modulus (82 GPa), the highest plastic strain (38%), high compressive yield strength (1078 MPa), hardness (334 Hv5) and elastic energy (7.08 MJ/m^3), which are better than those for commercial Ti-based biomaterials such as CP-Ti (552 MPa, 190 Hv, 1.13 MJ/m^3 respectively) and Ti-6Al-4V (970 MPa, 294 Hv, 3.10 MJ/m^3 respectively). Therefore, from the viewpoint of mechanical compatibility, the Ti-11Nb-9Fe alloy has a decent mix of mechanical properties for implant substitution application.

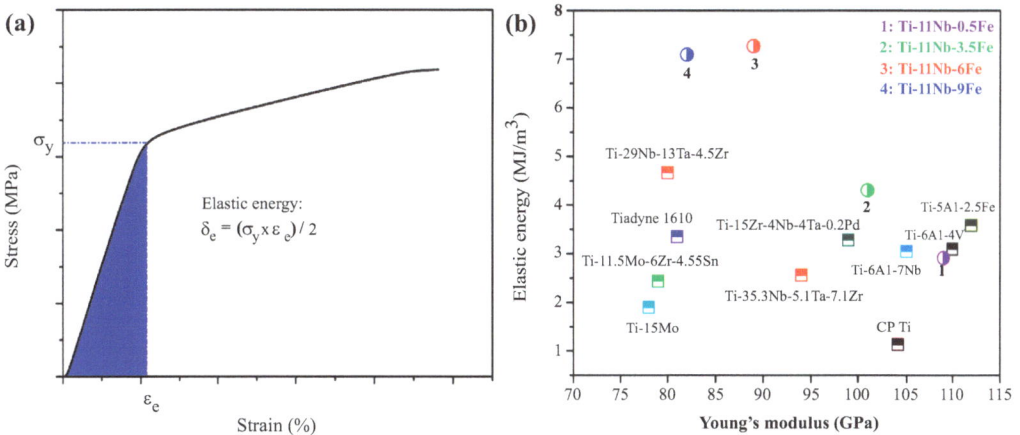

Fig. (7). (a) Elastic energy illustration, and **(b)** Elastic energy *versus* Young's modulus plot for studied alloys and some commercially used Ti-based alloys [45].

TI-FE BASED ALLOYS WITH MULTIPLE LENGTH-SCALE PHASES

The microstructure and mechanical properties of the aforementioned Ti-Fe based alloys, which were designed by using the DV-Xα molecular orbital method, are different from those in the literature [71]. Some examples of high strength Ti-Fe based alloys with multiple length-scale phases are shown for comparison. Zhang *et al.* [31] have investigated the microstructure, mechanical properties and deformation behavior of the hypereutectic $Ti_{65}Fe_{35}$ binary alloy by adding element Sn. The micrometer-sized FeTi primary dendrites are homogeneously dispersed in an ultrafine-grained matrix for $Ti_{65}Fe_{35}$ (Fig. **8a**) and $Ti_{63.375}Fe_{34.125}Sn_{2.5}$ (Fig. **8b**), which have shown a typical hypereutectic microstructure. This is revealed by a TEM micrograph shown on Fig. (**8d**): the plate-like FeTi (bright) and β-Ti (gray) phases are forms an ultra-fine eutectic matrix. As shown in Fig. (**8b**), the TiFe dendrite size decreases with adding 2.5% Sn to $Ti_{65}Fe_{35}$ alloy, and the grain size of the eutectic matrix in the alloy is improved. This is mainly reflected by the decreases in the width of the FeTi lamellae from ~400 nm to ~200 nm and in the width of β-Ti lamellae from the original ~2.5 µm reduced to ~160 nm. Also, the volume fraction of the ultrafine (β-Ti + FeTi) eutectic in the bimodal composites increases from ~65% to ~72%. The Energy-dispersive X-ray spectroscopy (EDS) analysis indicates that when 2.5% of Sn is added to the $Ti_{65}Fe_{35}$ alloy, most of the Sn are dissolved in the eutectic matrix. Such a type of solid solution hardening effect can make the alloy tougher. This result has been supported by the XRD results. As estimated from the XRD patterns, the presence of Sn in the TiFe alloy only changes the lattice parameter of β-Ti ($a_{β-Ti}$) and the lattice parameter of TiFe is remained the same as that in the original alloy (0.2991 nm). From the above analyses, it can be seen that the addition of 2.5% Sn to the $Ti_{65}Fe_{35}$ alloy refines the microstructure (such as slightly finer primary FeTi dendrites, prominent smaller sized eutectic matrix) together with the increased amount of the ductile matrix (~7 vol.%), thereby enhancing the plasticity and strength of the alloy. When the Sn content is increased from 2.5% to 5%, as shown in Fig. (**8c**), the microstructure shows a different phase constituents from $Ti_{63.375}Fe_{34.125}Sn_{2.5}$ alloy, which consists of Ti_3Sn (bright), FeTi (gray) and (β-Ti + FeTi) eutectic phase. However, the unchanged parameter is the grain size of TeFi phase (about 10 µm) and eutectic lamellar spacing between $Ti_{63.375}Fe_{34.125}Sn_{2.5}$ and $Ti_{61.75}Fe_{33.25}Sn_5$ alloy. From the above analyses, once can see that the phase structure, refinement and volume fraction of Ti-Fe-Sn ultrafine hypereutectic composite is affected by the Sn content added to the alloy. This may have an important influence on the strength and plasticity of ultrafine grained materials.

Table **3** summarizes the mechanical properties of all studied $(Ti_{0.65}Fe_{0.35})_{100-x}Sn_x$ (x = 0, 2.5 and 5%) alloys. As seen from the table, all the TiFeSn alloys exhibit considerable work hardening, high strength and large plasticity. When the Sn concentration is reaching 2.5%, the fracture strength increased significantly about 300 MPa, and the plastic is also increased by around 5%. Furthermore, when the Sn content increases to 5%, its strength and plasticity are slightly reduced, which is related to the new formed phase Ti_3Sn, but the plasticity is still higher than the original sample.

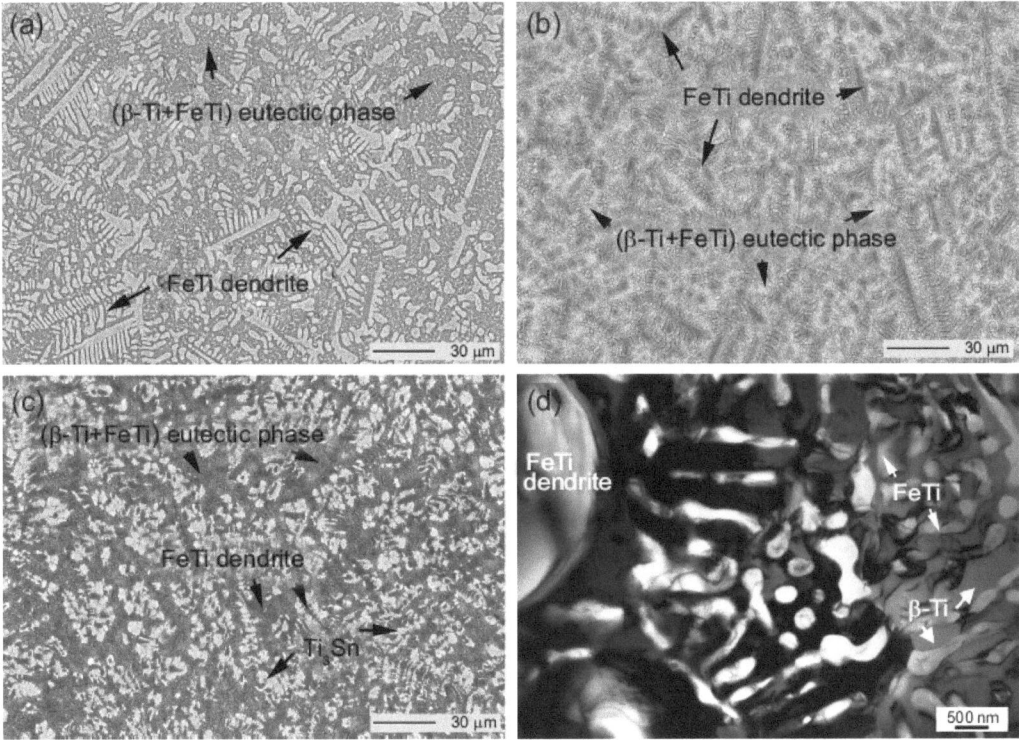

Fig. (8). Backscattered electron images for the $(Ti_{0.65}Fe_{0.35})_{100-x}Sn_x$ alloy: (**a**) x = 0, (**b**) x = 2.5 and (**c**) x = 5, and (**d**) bright-field TEM image of the $Ti_{63.375}Fe_{34.125}Sn_{2.5}$ [31, 72].

Table 3. The mechanical properties of $(Ti_{0.65}Fe_{0.35})_{100-x}Sn_x$ alloys: yield stress $\sigma_{0.2}$, ultimate stress σ_{max}, plastic strain ε_p and fracture strain ε_f [31].

Alloy nominal composition (at%)	Phase constituents	V_f (%)	$\sigma_{0.2}$ (MPa)	σ_{max} (MPa)	ε_p (%)	ε_f (%)
$Ti_{65}Fe_{35}$	FeTi, eutectic phase	36, 65	1722	2365	7.4	9.2
$Ti_{63.375}Fe_{34.125}Sn_{2.5}$	FeTi, eutectic phase	28, 72	1478	2652	12.5	14.5
$Ti_{61.75}Fe_{33.25}Sn_5$	Ti_3Sn, FeTi, eutectic phase	20, 24, 56	1267	2345	10.0	11.5

Fig. (9) summarizes the specific fracture strength of 3 types of typical engineering structural materials of nanostructured/ultrafine-crystalline bimodal composite materials, Ti-based bulk metallic glass (BMG) matrix composites and other commercial engineering structural materials. It is not difficult to see that the specific strength of the nanostructured/ultrafine crystal bimodal composite material is much higher than that of the commercial engineering structural material, and about 1.5 times more than that of the Ti-6Al-4V. This is due to its multiple length-size phases induced results. The Ti-based BMG matrix composites also exhibit similar results, although the specific strength is slightly lower than that of the nanostructured/ultrafine-crystalline bimodal composite. Therefore, due to the high strength and high specific strength of the nanostructured/ultrafine grain titanium alloy and the combination of large plasticity, it gives a higher potential application opportunities.

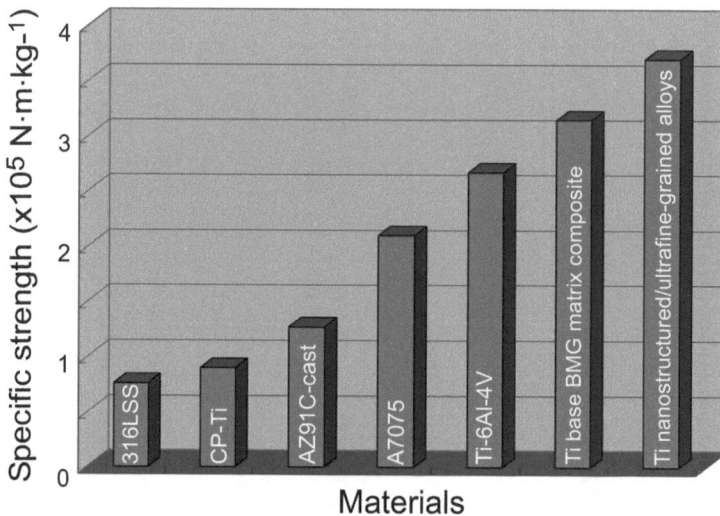

Fig. (9). Specific strength of nanostructured / ultrafine-grained titanium alloys with multiple length-scale phases and some other commercial engineering titanium alloys [71].

Of course, although the Ti-Fe based alloys have shown high strength and large plasticity due to the multiple length scale microstructure [71], extensive endeavors are still being made in developing high strength titanium alloys with large plasticity [73].

CONCLUDING REMARKS

This chapter reviews the recent developed Ti-Fe based titanium alloys in terms of microstructure and mechanical properties. Fe content influenced both microstructure and mechanical properties of Ti-11Nb-xFe alloys. The α phase-dominated Ti-11Nb-0.5Fe alloy presents the typical Widmanstätten structure. When Fe content increases to 3.5 wt%, the concentration of α phase decreases, while β phase quantity increases. At 6 wt% Fe, a significant amount of β phase is retained in the microstructure; however, a low fraction of α'' martensite is present. Finally, when Fe content is 9 wt%, the β phase is completely retained in the alloy microstructure. It is suggested that the stability of β phase in the aforementioned alloys is a function of their composition. Consequently, with increasing Fe content, the stability of β phase is enhanced, whereas the likelihood of the designed Ti-11Nb-xFe alloys to form α or α″ martensite upon cooling declines. The results obtained from compressive and hardness tests show that the ranges for compressive yield strength, hardness and elastic energy values of all alloys are 796-1137 MPa, 278-357 HV5 and 2.90-7.26 MJ/m³, respectively. These are higher than the corresponding ones of CP-Ti (552 MPa, 190 HV, 1.13 MJ/m3) or Ti-6Al-4V (970 MPa, 294 HV, 3.10 MJ/m³). Among all the Ti-11Nb-xFe alloys studied, Ti-11Nb-0.5Fe with the least β phase concentration, presents the highest Young's modulus (109 GPa) and the lowest plastic strain (18%), while β-type Ti-11Nb-9Fe alloy shows the greatest plastic strain (38%) and the lowest elastic modulus (82 GPa) which is much smaller than those of CP-Ti (104 GPa) and Ti-6Al-4V (114 GPa). It has been concluded that Ti-11Nb-9Fe alloy has the potential to be used for orthopaedic applications. However, it should be noted that by different application purpose, the alloy composition should be carefully designed.

CONFLICT OF INTEREST

The author (editor) declares no conflict of interest, financial or otherwise.

ACKNOWLEDGEMENTS

The author thanks Y.J. Liu, G.H. Cao, H.B. Lu, J. Eckert H. Attar, K.G. Prashanth A.K. Chaubey, and M. Calin for their collaborations.

REFERENCES

[1]   Hsu H-C, Hsu S-K, Wu S-C, Lee C-J, Ho W-F. Structure and mechanical properties of as-cast Ti-5Nb-xFe alloys. Mater Charact 2010; 61: 851-8.
[http://dx.doi.org/10.1016/j.matchar.2010.05.003]

[2]   Song B, Dong S, Zhang B, Liao H, Coddet C. Effects of processing parameters on microstructure and mechanical property of selective laser melted Ti-6Al-4V. Mater Des 2012; 35: 120-5.
[http://dx.doi.org/10.1016/j.matdes.2011.09.051]

[3]   Niinomi M, Nakai M, Hieda J. Development of new metallic alloys for biomedical applications. Acta Biomater 2012; 8(11): 3888-903.
[http://dx.doi.org/10.1016/j.actbio.2012.06.037] [PMID: 22765961]

[4]   Dai N, Zhang L-C, Zhang J, *et al.* Distinction in Corrosion Resistance of Selective Laser Melted Ti-6Al-4V Alloy on Different Planes. Corros Sci 2016; 111: 703-10.
[http://dx.doi.org/10.1016/j.corsci.2016.06.009]

[5]   Dai N, Zhang L-C, Zhang J, Chen Q, Wu M. Corrosion behavior of selective laser melted Ti-6Al-4V alloy in NaCl solution. Corros Sci 2016; 102: 484-9.
[http://dx.doi.org/10.1016/j.corsci.2015.10.041]

[6]   Chen Y, Zhang J, Dai N, Qin P, Attar H, Zhang LC. Corrosion Behaviour of Selective Laser Melted Ti-TiB Biocomposite in Simulated Body Fluid. Electrochim Acta 2017; 232: 89-97.
[http://dx.doi.org/10.1016/j.electacta.2017.02.112]

[7]   Dai N, Zhang J, Chen Y, Zhang LC. Heat Treatment Degrading the Corrosion Resistance of Selective Laser Melted Ti-6Al-4V Alloy. J Electrochem Soc 2017; 164: C428-34.

[http://dx.doi.org/10.1149/2.1481707jes]

[8]   Bai Y, Gai X, Li SJ, *et al.* Improved corrosion behaviour of electron beam melted Ti-6Al-4 V alloy in phosphate buffered saline. Corros Sci 2017; 123: 289-96.
[http://dx.doi.org/10.1016/j.corsci.2017.05.003]

[9]   Liu YJ, Li SJ, Hou WT, *et al.* Electron beam melted beta-type Ti-24Nb-4Zr-8Sn porous structures with high strength-to-modulus ratio. J Mater Sci Technol 2016; 32: 505-8.
[http://dx.doi.org/10.1016/j.jmst.2016.03.020]

[10]   Liu YJ, Li SJ, Wang HL, *et al.* Microstructure, defects and mechanical behavior of beta-type titanium porous structures manufactured by electron beam melting and selective laser melting. Acta Mater 2016; 113: 56-67.
[http://dx.doi.org/10.1016/j.actamat.2016.04.029]

[11]   Liu YJ, Li XP, Zhang LC, Sercombe TB. Processing and properties of topologically optimised biomedical Ti-24Nb-4Zr-8Sn scaffolds manufactured by selective laser melting. Mater Sci Eng A 2015; 642: 268-78.
[http://dx.doi.org/10.1016/j.msea.2015.06.088]

[12]   Liu YJ, Wang HL, Li SJ, *et al.* Compressive and fatigue behavior of beta-type titanium porous structures fabricated by electron beam melting. Acta Mater 2017; 126: 58-66.
[http://dx.doi.org/10.1016/j.actamat.2016.12.052]

[13]   Yang Y, Li GP, Wang H, *et al.* Formation of zigzag-shaped {112} <111> β mechanical twins in Ti-24.5 Nb-0.7 Ta-2 Zr-1.4 O alloy. Scr Mater 2012; 66: 211-4.
[http://dx.doi.org/10.1016/j.scriptamat.2011.10.031]

[14]   Wang L, Qu J, Chen L, *et al.* Investigation of Deformation Mechanisms in β-Type Ti-35Nb-2Ta-3Zr Alloy *via* FSP Leading to Surface Strengthening. Metall Mater Trans, A Phys Metall Mater Sci 2015; 46: 4813-8.
[http://dx.doi.org/10.1007/s11661-015-3089-8]

[15]   Wang L, Xie L, Lv Y, *et al.* Microstructure evolution and superelastic behavior in Ti-35Nb-2Ta-3Zr alloy processed by friction stir processing. Acta Mater 2017; 131: 499-510.
[http://dx.doi.org/10.1016/j.actamat.2017.03.079]

[16]   Liu LH, Yang C, Wang F, *et al.* Ultrafine grained Ti-based composites with ultrahigh strength and ductility achieved by equiaxing microstructure. Mater Des 2015; 79: 1-5.
[http://dx.doi.org/10.1016/j.matdes.2015.04.032]

[17]   Ho W-F, Pan C-H, Wu S-C, Hsu H-C. Mechanical properties and deformation behavior of Ti-5Cr-xFe alloys. J Alloys Compd 2009; 472: 546-50.
[http://dx.doi.org/10.1016/j.jallcom.2008.05.015]

[18]   Kuroda D, Niinomi M, Morinaga M, Kato Y, Yashiro T. Design and mechanical properties of new β type titanium alloys for implant materials. Mater Sci Eng A 1998; 243: 244-9.
[http://dx.doi.org/10.1016/S0921-5093(97)00808-3]

[19]   Abdel-Hady Gepreel M, Niinomi M. Biocompatibility of Ti-alloys for long-term implantation. J Mech Behav Biomed Mater 2013; 20: 407-15.
[http://dx.doi.org/10.1016/j.jmbbm.2012.11.014] [PMID: 23507261]

[20]   Milošev I, Žerjav G, Moreno JM, Popa M. Electrochemical properties, chemical composition and thickness of passive film formed on novel Ti-20Nb-10Zr-5Ta alloy. Electrochim Acta 2013; 99: 176-89.
[http://dx.doi.org/10.1016/j.electacta.2013.03.086]

[21]   Li Y, Yang C, Zhao H, Qu S, Li X, Li Y. New Developments of Ti-Based Alloys for Biomedical Applications. Materials (Basel) 2014; 7(3): 1709-800.
[http://dx.doi.org/10.3390/ma7031709] [PMID: 28788539]

[22]   Li Y, Zou L, Yang C, Li Y, Li L. Ultrafine-grained Ti-based composites with high strength and low

modulus fabricated by spark plasma sintering. Mater Sci Eng A 2013; 560: 857-61.
[http://dx.doi.org/10.1016/j.msea.2012.09.047]

[23]   Zhang Y, Kent D, Wang G, St John D, Dargusch M. An investigation of the mechanical behaviour of fine tubes fabricated from a Ti-25Nb-3Mo-3Zr-2Sn alloy. Mater Des 2015; 85: 256-65.
[http://dx.doi.org/10.1016/j.matdes.2015.06.127]

[24]   Haghighi Ehtemam S, Lu HB, Jian GY, Cao GH, Habibi D, Zhang LC. Effect of α" martensite on the microstructure and mechanical properties of beta-type Ti-Fe-Ta alloys. Mater Des 2015; 76: 47-54.
[http://dx.doi.org/10.1016/j.matdes.2015.03.028]

[25]   Abdel-Hady M, Hinoshita K, Morinaga M. General approach to phase stability and elastic properties of β-type Ti-alloys using electronic parameters. Scr Mater 2006; 55: 477-80.
[http://dx.doi.org/10.1016/j.scriptamat.2006.04.022]

[26]   Collings E. The physical metallurgy of titanium alloys. Ohio: ASM International 1984.

[27]   Kim H, Hashimoto S, Kim J, Inamura T, Hosoda H, Miyazaki S. Effect of Ta addition on shape memory behavior of Ti-22Nb alloy. Mater Sci Eng A 2006; 417: 120-8.
[http://dx.doi.org/10.1016/j.msea.2005.10.065]

[28]   Li C, Zhan Y, Jiang W. β-Type Ti-Mo-Si ternary alloys designed for biomedical applications. Mater Des 2012; 34: 479-82.
[http://dx.doi.org/10.1016/j.matdes.2011.08.012]

[29]   Mantani Y, Tajima M. Phase transformation of quenched α″ martensite by aging in Ti-Nb alloys. Mater Sci Eng A 2006; 438: 315-9.
[http://dx.doi.org/10.1016/j.msea.2006.02.180]

[30]   Cui C, Ping D. Microstructural evolution and ductility improvement of a Ti-30Nb alloy with Pd addition. J Alloys Compd 2009; 471: 248-52.
[http://dx.doi.org/10.1016/j.jallcom.2008.03.057]

[31]   Zhang LC, Das J, Lu HB, Duhamel C, Calin M, Eckert J. High strength Ti-Fe-Sn ultrafine composites with large plasticity. Scr Mater 2007; 57: 101-4.
[http://dx.doi.org/10.1016/j.scriptamat.2007.03.031]

[32]   Zhang LC, Lu HB, Mickel C, Eckert J. Ductile ultrafine-grained Ti-based alloys with high yield strength. Appl Phys Lett 2007; 91: 051906.
[http://dx.doi.org/10.1063/1.2766861]

[33]   Xu L, Chen Y, Liu ZG, Kong F. The microstructure and properties of Ti-Mo-Nb alloys for biomedical application. J Alloys Compd 2008; 453: 320-4.
[http://dx.doi.org/10.1016/j.jallcom.2006.11.144]

[34]   Lin DJ, Lin JH, Ju C-P. Structure and properties of Ti-7.5Mo-xFe alloys. Biomaterials 2002; 23(8): 1723-30.
[http://dx.doi.org/10.1016/S0142-9612(01)00233-2] [PMID: 11950042]

[35]   Majumdar P, Singh S, Chakraborty M. Elastic modulus of biomedical titanium alloys by nano-indentation and ultrasonic techniques-A comparative study. Mater Sci Eng A 2008; 489: 419-25.
[http://dx.doi.org/10.1016/j.msea.2007.12.029]

[36]   Gabriel S, Panaino J, Santos I, *et al.* Characterization of a new beta titanium alloy, Ti-12Mo-3Nb, for biomedical applications. J Alloys Compd 2012; 536: S208-10.
[http://dx.doi.org/10.1016/j.jallcom.2011.11.035]

[37]   Chen YY, Xu LJ, Liu ZG, Kong FT, Chen ZY. Microstructures and properties of titanium alloys Ti-Mo for dental use. Trans Nonferrous Met Soc China 2006; 16: s824-28.
[http://dx.doi.org/10.1016/S1003-6326(06)60308-7]

[38]   Calin M, Helth A, Gutierrez Moreno JJ, *et al.* Elastic softening of β-type Ti-Nb alloys by indium (In) additions. J Mech Behav Biomed Mater 2014; 39: 162-74.

[http://dx.doi.org/10.1016/j.jmbbm.2014.07.010] [PMID: 25128870]

[39] Lu J, Zhao Y, Niu H, *et al.* Electrochemical corrosion behavior and elasticity properties of Ti-6Al-xFe alloys for biomedical applications. Mater Sci Eng C 2016; 62: 36-44.
[http://dx.doi.org/10.1016/j.msec.2016.01.019] [PMID: 26952395]

[40] Kuroda D, Kawasaki H, Yamamoto A, Hiromoto S, Hanawa T. Mechanical properties and microstructures of new Ti-Fe-Ta and Ti-Fe-Ta-Zr system alloys. Mater Sci Eng C 2005; 25: 312-20.
[http://dx.doi.org/10.1016/j.msec.2005.04.004]

[41] Abd-elrhman Y, Gepreel MA, Abdel-Moniem A, Kobayashi S. Compatibility assessment of new V-free low-cost Ti-4.7 Mo-4.5 Fe alloy for some biomedical applications. Mater Des 2016; 97: 445-53.
[http://dx.doi.org/10.1016/j.matdes.2016.02.110]

[42] Nouri A, Hodgson PD, Wen C. Effect of ball-milling time on the structural characteristics of biomedical porous Ti-Sn-Nb alloy. Mater Sci Eng C 2011; 31: 921-8.
[http://dx.doi.org/10.1016/j.msec.2011.02.011]

[43] Kopova I, Stráský J, Harcuba P, Landa M, Janeček M, Bačákova L. Newly developed Ti-Nb-Zr-T--Si-Fe biomedical beta titanium alloys with increased strength and enhanced biocompatibility. Mater Sci Eng C 2016; 60: 230-8.
[http://dx.doi.org/10.1016/j.msec.2015.11.043] [PMID: 26706526]

[44] Ehtemam-Haghighi S, Cao G, Zhang LC. Nanoindentation study of mechanical properties of Ti based alloys with Fe and Ta additions. J Alloys Compd 2017; 692: 892-7.
[http://dx.doi.org/10.1016/j.jallcom.2016.09.123]

[45] Ehtemam-Haghighi S, Liu Y, Cao G, Zhang L-C. Phase Transition, Microstructural Evolution And Mechanical Properties Of Ti-Nb-Fe Alloys Induced By Fe Addition. Mater Des 2016; 97: 279-86.
[http://dx.doi.org/10.1016/j.matdes.2016.02.094]

[46] Ehtemam-Haghighi S, Liu Y, Cao G, Zhang LC. Influence of Nb on the $\beta \rightarrow \alpha''$ martensitic phase transformation and properties of the newly designed Ti-Fe-Nb alloys. Mater Sci Eng C 2016; 60: 503-10.
[http://dx.doi.org/10.1016/j.msec.2015.11.072] [PMID: 26706557]

[47] Morinaga M, Yukawa N, Maya T, Sone K, Adachi H, Eds. Theoretical design of titanium alloys. Sixth World Conference on Titanium III.

[48] Min X, Emura S, Sekido N, Nishimura T, Tsuchiya K, Tsuzaki K. Effects of Fe addition on tensile deformation mode and crevice corrosion resistance in Ti-15Mo alloy. Mater Sci Eng A 2010; 527: 2693-701.
[http://dx.doi.org/10.1016/j.msea.2009.12.050]

[49] Xu W, Kim K, Das J, Calin M, Eckert J. Phase stability and its effect on the deformation behavior of Ti-Nb-Ta-In/Cr β alloys. Scr Mater 2006; 54: 1943-8.
[http://dx.doi.org/10.1016/j.scriptamat.2006.02.002]

[50] Lopes E, Cremasco A, Afonso C, Caram R. Effects of double aging heat treatment on the microstructure, Vickers hardness and elastic modulus of Ti-Nb alloys. Mater Charact 2011; 62: 673-80.
[http://dx.doi.org/10.1016/j.matchar.2011.04.015]

[51] Gabriel SB, Dille J, Nunes CA. Soares GdA. The effect of niobium content on the hardness and elastic modulus of heat-treated Ti-10Mo-xNb alloys. Mater Res 2010; 13: 333-7.
[http://dx.doi.org/10.1590/S1516-14392010000300009]

[52] Welsch G, Boyer R, Collings E. Materials properties handbook: titanium alloys. ASM International 1993.

[53] Yan W, Berthe J, Wen C. Numerical investigation of the effect of porous titanium femoral prosthesis on bone remodeling. Mater Des 2011; 32: 1776-82.
[http://dx.doi.org/10.1016/j.matdes.2010.12.042]

[54]   Ren Y, Wang F, Wang S, *et al.* Mechanical response and effects of β-to-α" phase transformation on the strengthening of Ti-10V-2Fe-3Al during one-dimensional shock loading. Mater Sci Eng A 2013; 562: 137-43.
[http://dx.doi.org/10.1016/j.msea.2012.10.098]

[55]   Calin M, Gebert A, Ghinea AC, *et al.* Designing biocompatible Ti-based metallic glasses for implant applications. Mater Sci Eng C 2013; 33(2): 875-83.
[http://dx.doi.org/10.1016/j.msec.2012.11.015] [PMID: 25427501]

[56]   He G, Eckert J, Dai QL, *et al.* Nanostructured Ti-based multi-component alloys with potential for biomedical applications. Biomaterials 2003; 24(28): 5115-20.
[http://dx.doi.org/10.1016/S0142-9612(03)00440-X] [PMID: 14568427]

[57]   Zhu Y, Wang X, Wang L, *et al.* Influence of forging deformation and heat treatment on microstructure of Ti-xNb-3Zr-2Ta alloys. Mater Sci Eng C 2012; 32: 126-32.
[http://dx.doi.org/10.1016/j.msec.2011.10.006]

[58]   Zhou YL, Niinomi M, Akahori T. Effects of Ta content on Young's modulus and tensile properties of binary Ti-Ta alloys for biomedical applications. Mater Sci Eng A 2004; 371: 283-90.
[http://dx.doi.org/10.1016/j.msea.2003.12.011]

[59]   Ehtemam-Haghighi S, Prashanth KG, Attar H, Chaubey AK, Cao GH, Zhang LC. Evaluation Of Mechanical And Wear Properties Of Tixnb7fe Alloys Designed For Biomedical Applications. Mater Des 2016; 111: 592-9.
[http://dx.doi.org/10.1016/j.matdes.2016.09.029]

[60]   Attar H, Prashanth K, Chaubey A, *et al.* Comparison of wear properties of commercially pure titanium prepared by selective laser melting and casting processes. Mater Lett 2015; 142: 38-41.
[http://dx.doi.org/10.1016/j.matlet.2014.11.156]

[61]   Geetha M, Singh A, Asokamani R, Gogia A. Ti based biomaterials, the ultimate choice for orthopaedic implants-a review. Prog Mater Sci 2009; 54: 397-425.
[http://dx.doi.org/10.1016/j.pmatsci.2008.06.004]

[62]   Peters M, Hemptenmacher J, Kumpfert J, Leyens C. Structure and properties of titanium and titanium alloys.Titanium and titanium alloys: fundamentals and applications. Weinheim: WILEY-VCH 2003; pp. 1-36.
[http://dx.doi.org/10.1002/3527602119.ch1]

[63]   Faria AC, Rodrigues RC, Claro AP, da Gloria Chiarello de Mattos M, Ribeiro RF. Wear resistance of experimental titanium alloys for dental applications. J Mech Behav Biomed Mater 2011; 4(8): 1873-9.
[http://dx.doi.org/10.1016/j.jmbbm.2011.06.004] [PMID: 22098886]

[64]   Carman A, Zhang LC, Ivasishin O, Savvakin D, Matviychuk M, Pereloma E. Role of alloying elements in microstructure evolution and alloying elements behaviour during sintering of a near-β titanium alloy. Mater Sci Eng A 2011; 528: 1686-93.
[http://dx.doi.org/10.1016/j.msea.2010.11.004]

[65]   Min X, Zhang L, Sekido K, *et al.* Strength evaluation of α and β phases by nanoindentation in Ti-15Mo alloys with Fe and Al addition. Mater Sci Technol 2012; 28: 342-7.
[http://dx.doi.org/10.1179/1743284711Y.0000000052]

[66]   Kolli RP, Joost WJ, Ankem S. Phase stability and stress-induced transformations in beta titanium alloys. J Miner Met Mater Soc 2015; 67: 1273-80.
[http://dx.doi.org/10.1007/s11837-015-1411-y]

[67]   Niinomi M. Mechanical properties of biomedical titanium alloys. Mater Sci Eng A 1998; 243: 231-6.
[http://dx.doi.org/10.1016/S0921-5093(97)00806-X]

[68]   Ho W-F, Cheng C-H, Pan C-H, Wu S-C, Hsu H-C. Structure, mechanical properties and grindability of dental Ti-10Zr-X alloys. Mater Sci Eng C 2009; 29: 36-43.
[http://dx.doi.org/10.1016/j.msec.2008.05.004]

[69]   Zhang D, Mao Y, Li Y, Li J, Yuan M, Lin J. Effect of ternary alloying elements on microstructure and superelastictity of Ti-Nb alloys. Mater Sci Eng A 2013; 559: 706-10.
[http://dx.doi.org/10.1016/j.msea.2012.09.012]

[70]   Zhan Y, Li C, Jiang W. β-type Ti-10Mo-1.25Si-xZr biomaterials for applications in hard tissue replacements. Mater Sci Eng C 2012; 32(6): 1664-8.
[http://dx.doi.org/10.1016/j.msec.2012.04.059] [PMID: 24364974]

[71]   Zhang LC. High performance ultrafine-grained Ti-Fe-based alloys with multiple length-scale phases. Adv Mat Res 2012; 1: 13-29.
[http://dx.doi.org/10.12989/amr.2012.1.1.013]

[72]   Zhang LC, Lu HB, Calin M, Pereloma EV, Eckert J. High-strength ultrafine-grained Ti-Fe-Sn alloys with a bimodal structure. J Phys Conf Ser 2010; 240: 012103.
[http://dx.doi.org/10.1088/1742-6596/240/1/012103]

[73]   Yang C, Kang LM, Li XX, *et al.* Bimodal titanium alloys with ultrafine lamellar eutectic structure fabricated by semi-solid sintering. Acta Mater 2017; 132: 491-502.
[http://dx.doi.org/10.1016/j.actamat.2017.04.062]

CHAPTER 4

Selective Laser Melting of Titanium Alloys: Processing, Microstructure and Properties

Lai-Chang Zhang[1,*] and **Liqiang Wang**[2]

[1] *School of Engineering, Edith Cowan University, Perth, WA, Australia*

[2] *State Key Laboratory of Metal Matrix Composites, Shanghai Jiao Tong University, Shanghai, China*

Abstract: Although titanium alloys have exhibited a combination range of excellent properties. However, various potential applications of titanium alloys are hampered by hard machinery and/or high cost due to material removal in conventional manufacturing processes. Emerging additive manufacturing techniques are providing a perfect opportunity for creating titanium and its composites, especially with complex dimensions, such as selective laser melting (SLM). So far, many types of titanium alloys components have been successfully manufactured by SLM, and they show comparable properties compared with their traditional counterparts. This chapter first briefly introduces the characteristics of the SLM process and parameters involved, then reviews some of the latest developments in the processing, microstructure and mechanical properties of Ti-based alloys and porous structures produced by SLM.

Keywords: Composites, Deformation behavior, Fatigue, Fracture, Mechanical properties, Microstructure, Selective laser melting, Titanium, Wear.

INTRODUCTION

The properties of titanium, such as the combination of low density (4500 kg/m^3), favorable mechanical properties, high specific strength, satisfying biocompatibility and excellent corrosion resistance, makes titanium alloys become one of the superior engineering materials, and suitable for many industrial applications [1 - 18]. For example, the Young's modulus of titanium alloys in the biomaterial study have a much lower value (55-110 GPa) than other commonly used biomaterials alloys like Co-Cr alloys (240 GPa) and 316L stainless steel (210 GPa). Therefore, the titanium alloys are perfect for hard tissue implants application. The traditional processing technology to produce the titanium alloys is solidification or casting [5, 14, 19 - 24], powder metallurgy [25 - 27], space

* **Corresponding author Lai-Chang Zhang:** School of Engineering, Edith Cowan University, 270 Joondalup Drive, Joondalup, Perth, WA 6027, Australia; Tel: 61 8 63042322; Fax: 61 8 63045811; E-mails: lczhangimr@gmail.com; l.zhang@ecu.edu.au

holder technology [28] and sheet forming [29]. In general, these traditional techniques for producing the Ti-based alloys involves typically massive manufacturing time, more material waste and large energy consuming. The causes for these drawbacks are related to the high reactivity of titanium, the complicated extraction process and the difficulty of melting due to the high melting points. Those negative features will lead to the problems such as the cost of manufacturing. Therefore, mechanization for the titanium alloys are much higher than other metal materials, thereby significantly hindering the potential application of titanium alloys. Apparently, precision or near-net shape machining is desirable for making titanium components.

Selective laser melting (SLM) and electron beam melting (EBM), as the new metallic additive manufacturing technologies, are giving a perfect platform to the generation of titanium components [30, 31]. This sort of innovation can coordinate utilization of metal powder materials with computer-aided design (CAD) model to deliver a scope of imperative engineering metal compounds with near full density. The SLM technique offers an extensive variety of favorable properties in the examination with the traditional standard manufacturing techniques. For example, precision machining generation, short creation cycle, high material utilization and almost no geometric constriction. These advantages explicitly prompting create complex-shape parts virtually without additionally post-processing [32 - 35]. SLM is the best alternative method for the production of titanium alloy composites with complex shapes, due to its manufacturing process and core processing technology. A high degree of evaluation of the SLM-produced product has led to a great deal of concern in the manufacturing of various metallic parts, which including titanium [34 - 36], aluminium [32, 33, 37 - 40], steels [41, 42], superalloy [43, 44] and bronze materials [45]. Compared with the relatively limited applications of EBM in titanium alloys [46 - 49], SLM has been applied to manufacture all types of titanium alloys, like the titanium matrix composites, titanium alloys with different phases (α / β / $\alpha+\beta$ type) and porous structures. In this chapter, SLM process and its processing parameters are first briefly introduced. Afterwards, the recent progress of SLM-produced titanium alloys is discussed with focus on processing, microstructure and properties.

SLM CHARACTERISTICS AND PROCESSING PARAMETERS

Process Overview

The primary process of conventional subtractive manufacturing to get a specific shape is removing the defective layers/parts from the original materials so that the subtractive manufacturing process is also called a material removal process. The SLM manufacturing method is different from the traditional subtractive manufacturing process. This technique, through the computer controlled laser beam, selectively melts and consolidates the starting powder materials layer by layer under the protective inert Argon gas. Although there already exist many different types of SLM equipment on the market, an SLM system usually includes a computer control system, a processing laser, an automatic powder feeder and some major ancillary components such as inert gas system protection device, drum/scrapper and overflow container [50]. The SLM process is schematically illustrated in Fig. (**1**) and involves the following steps.

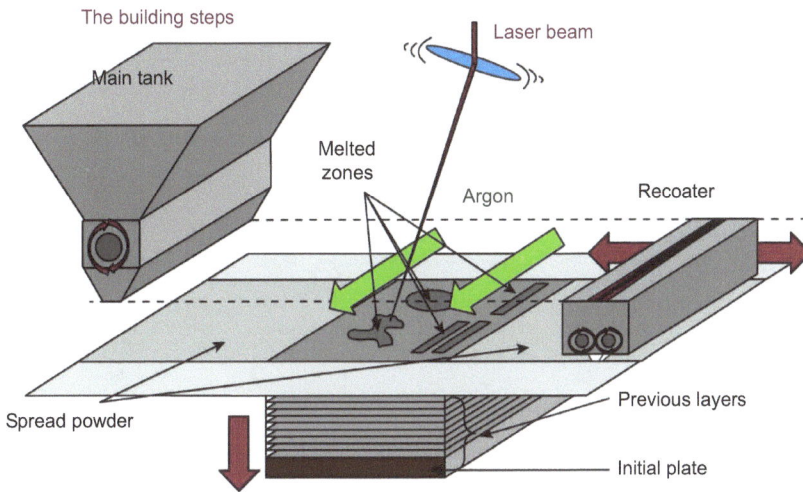

Fig. (1). Schematic illustration of the selective laser melting process [51].

1. *3D modeling:* Use the CAD software package to model the product to be constructed. The designer can use pre-existing CAD files or may create a new model (the product to be fabricated).
2. *Model conversion:* Covent the file from the CAD model to STL format, as the STL (stereolithography) is the standard format for all additive manufacturing technologies.
3. *Slicing and finalizing the model processing:* Mathematically slicing the STL model into some typically 45 μm thick layers, for preparing the STL file to be built. In the build file, the part as well as supports have been sliced. All the processing parameters are set in the file for fabrication.
4. *Materialized model:* After settling leveling the substrate on the build platform of the particular SLM device, a thin layer of the loose powder is deposited on the building substrate with the same thickness as the slice layer. Then, use the pre-set scan parameters and designed model to scan and deal with the powder bed. After creation of the first layer, the remaining layer repeats the same manufacturing process until the entire assembly is complete [50].

5. *Clean and finish:* This is the post-processing involves removing the manufactured builds from the platform of the machine and detaching any supports.

Processing Parameters Involved

The SLM process involves a large number of parameters as shown in Fig. (**2**), which include fixed parameters and processing parameters. The fixed parameters are determined by the SLM device itself (such as laser wavelength and laser working mode) and the material itself (like viscosity and heat conductivity), which are not admitted changing. The remaining parameters, called processing parameters, should be developed for a given material and be optimized in SLM process to achieve high-quality products [35, 52 - 54].

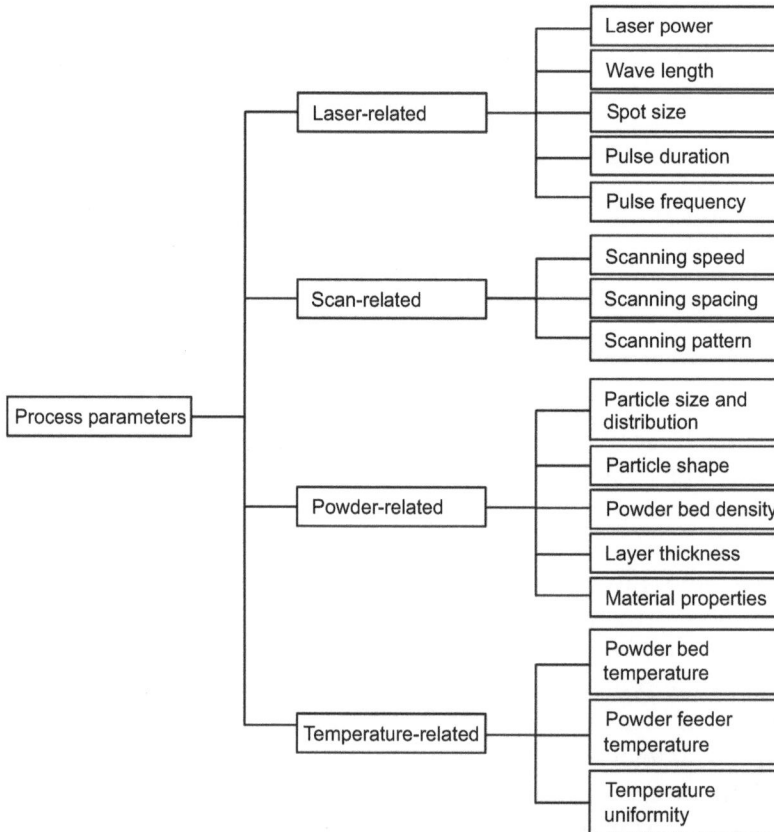

Fig. (2). The principal parameters involved in SLM process [55].

The utmost objective of SLM is to fabricate full density parts and without deformities to accomplish the coveted performance to catch up with the composites created by the traditional producing methods. The densification, quality and the properties of SLM-fabricated parts are closely related to laser energy density (*E*) applied during the SLM processing, for a given material, which is generally defined by [34]:

$$E = P/(v \cdot t \cdot s) \tag{1}$$

where the formula can be described as the laser power P (W) divided by the product of the scan speed v (mm/s), layer thickness t (mm) and scan spacing s (mm). As seen from Eq. **(1)**, in order to increase the powder melting temperature, a higher laser energy density is required. So, to increase the laser energy density is corresponding to enhance the laser power, and/or reducing any/all of the following parameters: scan speed, layer thickness, scan spacing.

Generally, a higher laser energy density produces a higher melting temperature of the powder. Therefore, the more melt will present, which enhances the density of the part [34, 35, 52 - 54, 56]. Since the ultimate goal of obtaining a higher level of metal density is recognized, and a significant amount of melt needs a higher incident laser energy density. Because the complete melting is the basis for the manufacture of full dense components, so obtaining a high dense composite requires to applying a sufficient laser energy density to the powder material. For example, to manufacture a fully dense part by using SLM, the critical laser energy density is around 120 J/mm^3 for commercially pure titanium (CP-Ti) [53, 54] and Ti-6Al-4V [56], and 40 J/mm^3 for Ti-24Nb-4Zr-8Sn [52]. However, during the SLM production, with the sufficient laser energy density reaching the maximum density of as-built parts, the scan speed, layer thickness and scan spacing should be carefully selected to ensure high-quality parts. For example, surface roughness and short production time from an economic point of view should be simulataeneous considered.

Scanning strategies is another crucial parameter that affects SLM manufacturing products, the laser scanning vectors which include length and pattern. The length of scanning vector is characterized by scanning geometry. This scanning, known as "scanning pattern", can be designed in various ways. The laser scanning mode is usually parallel to the straight line, but there may also be circular or spiral coverage. The scanning direction can be changed between the single layer or continuous layer [30]. The scanning direction can be pivoted from the interval between the island segments or layers, which are similar to the island scanning (inter-layer) or between sequential layers by various rotation angles. The design of this manufacturing method affects the quality of the SLM-processed components and therefore the performance of the produced parts.

SLM OF SOLID TITANIUM ALLOYS

SLM of α-type CP-Ti

The studies on SLM of CP-Ti, show that the processing parameters have significant effects on the densification of CP-Ti, which are mainly manifested in the microstructure, mechanical and wear properties [36, 54, 57, 58]. In most cases, the microstructure and mechanical properties of the SLM-produced samples are firmly identified to be closely related to the SLM processing parameters. As seen in Fig. (3), the SLM processing parameters scan speed and laser power are the two main factors that significantly affect the relative density of SLM-produced samples. Remarkably variable density is observed at the same level of energy density due to (1) incomplete melting of powder, (2) instability of the melt pool, (3) overheating of the powder, and (4) balling effects caused by some combinations of laser power and laser scan speed [54]. At the same level of laser energy density of 120 J/mm^3, the density of the SLM-produced part increases with the enhance in laser power and laser scan speed, and reaches a maximum value (about 99.5%) at the laser power of 165W [36, 54]. However, if further increasing the laser power over 165W, the density of the product is slightly reduced, which also indicates that the density is not exactly proportional to the laser power. In the combination of higher laser power and scan speed, the porosity is smaller and closer to the circle shape; this is due to the balling effect and high thermal stress cracking. In contrast, the porosity produced by the combination of lower laser power and lower scan speed is filled with un-melted powder because of insufficient melting of the metal powder. This has been proved by the presence of the balling effect in a sample produced at a low laser power and low scan speed [54].

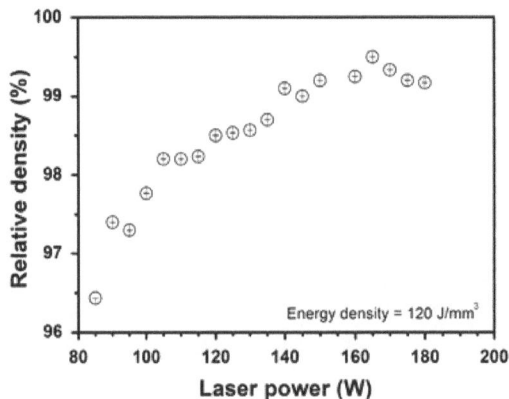

Fig. (3). Impact of laser power control on the relative density of SLM-produced CP-Ti parts at a constant energy density of 120 J/mm^3 [54].

The processing parameters also can alter the microstructure of SLM-produced components, such as applying a constant laser energy density of 120 J/mm^3, the plate-like α phase of CP-Ti samples changes to acicular martensitic α′ phase (Fig. **4**) by altering the laser scan speed in SLM [36]. When the laser scan speed is less than 100 mm/s, the plate-like α phase is formed during the solidification process, the energy thermalization within melt pool leads to a complete isomorphic transformation from β phase to α phase (Fig. **4a**). However, when the laser scan speed is higher than 100 mm/s, it will enhance the kinetic as well as the thermal characteristics under cooling, thereby increasing temperature gradients within the melt pool. This phenomenon results in formation of the acicular martensitic α′ phase in the SLM-produced CP-Ti part (Fig. **4b**). By comparison, the acicular α′ phase combined with grain refinement formed by high cooling significantly enhance the mechanical properties of SLM-produced CP-Ti [54]. The SLM-produced CP-Ti exhibits a yield strength of 555 MPa and an ultimate tensile strength of 757 MPa, better than the corresponding properties under sheet forming and full annealed counterparts, and without a noticeable diminishment in ductility. Also, the SLM-produced CP-Ti composite demonstrates higher Vickers microhardness (261 Hv [54]) than the cast alloys (210 Hv [59]), comparable to that for 55% cold rolled CP-Ti (268 Hv [60]).

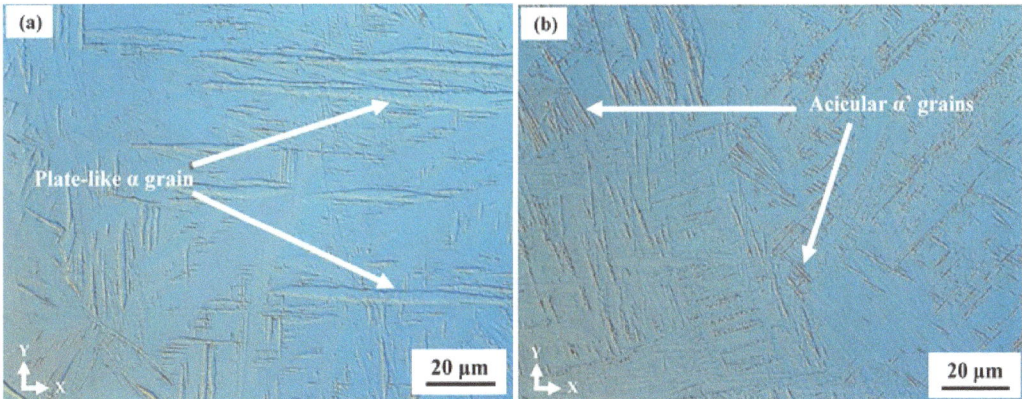

Fig. (4). Optical microstructure of the SLM-produced CP-Ti samples: (**a**) Plate-like α phase and (**b**) acicular α′ phase [54].

Furthermore, the wear study on the SLM-produced and the cast CP-Ti samples [61] demonstrates that because of the high cooling rate of SLM processing, the produced samples behave grain refinement, martensitic microstructure and high microhardness. This can directly lead to the CP-Ti samples with lower wear rate, *i.e.* better wear resistance (Fig. **5**). However, the abrasive component and oxidative wear mechanisms of SLM produced CP-Ti and cast CP-Ti are similar. The abrasive component occupies the plowing grooves of the wear, and the surface cracks can cause the surface to delaminate, resulting in the removal of the material.

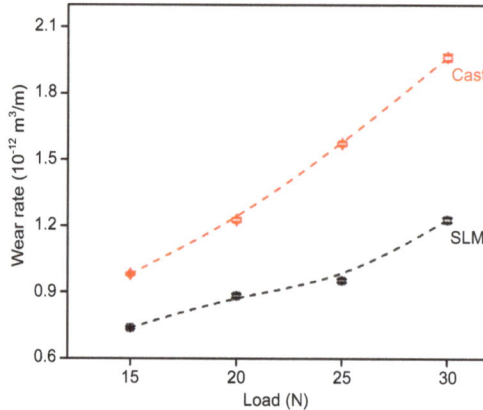

Fig. (5). Comparison of sliding wear rates of SLM-produced and cast CP-Ti as a function of load [61].

For further improving the poor hardness and wear properties of CP-Ti, SLM produced titanium matrix composites such as Ti-TiC and Ti-TiB composites have also been studied [53, 56, 62]. In the event that no titanium matrix composites powder is accessible due to the lack storage for SLM in this situation, 2-hour ball milling can create titanium matrix composites powder with a near-spherical shape [63]. The Ti-TiB composites with particles with needle-like morphology distributing homogeneously in α matrix can be obtained by SLM of the ball-milled composite powder as shown in Fig. (6) [53]. The SLM-produced the Ti-TiB composite also exhibits a higher microhardness compared to CP-Ti and Ti-6Al-4V, which can be identified with the hardening effect resulting from TiB particles and grain refinement of α-Ti. Moreover, compared with CP-Ti and Ti-6Al-4V, the SLM-fabricated Ti-TiB composite exhibits higher yield strength and higher ultimate strength in compression testing, which result from the strengthening effect of TiB particles. For Ti-TiC composites produced by SLM, the Ti-TiC composite exhibits approximately 22.7 times dynamic stiffness and about 2.4 times the elastic modulus of the non-reinforced titanium part of due to the uniform dispersion enhancement effect of the nanostructured TiC reinforcement [56].

SLM of (α+β)-type Titanium Alloys

Due to the easy availability of the powder feedstock of Ti-6Al-4V, (α+β)-type Ti-6Al-4V has been widely used in SLM research, especially in densification behavior, microstructure and mechanical properties [34, 64, 65]. The microstructural observations show that the SLM-produced Ti-6Al-4V sample exhibits a substantial amount of acicular α′ martensite together with some prior β grains (as shown in Fig. **7**) [66 - 69]. This microstructural feature is different from the typical α+β microstructure in both traditional Grade 5 sample [67] and EBM-produced Ti-6Al-4V [70]. However, the replacement of the equilibrium α+β phase by this metastable α′+β phase is very common in samples produced by SLM. Since the formation of α' martensite mentioned above has a relationship with the scan speed of SLM, faster scanning rates increase the kinetic energy and thermal characteristics during cooling, which can increase the temperature gradients within the melt pool. So the formation of α' martensite is attributed to the cooling rate of Ti-6Al-4V alloy produced by SLM (10^3-10^8K/s [71]), which is much higher than the critical cooling rate of 410K/s for α martensitic transformation from β phase, and a substantial thermal gradient (10^4-10^5 K/cm) along the build direction in tiny melt pool.

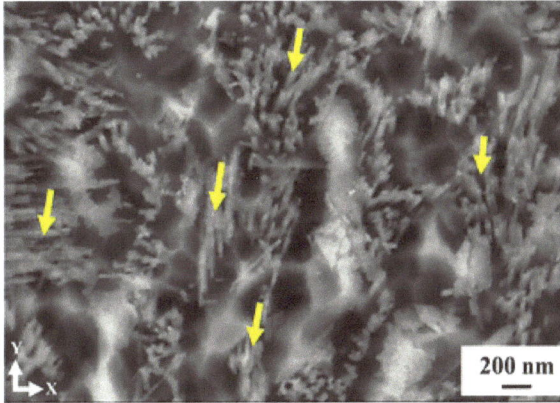

Fig. (6). SEM image of the SLM-produced Ti-TiB composite (TiB particles are arrowed) [53].

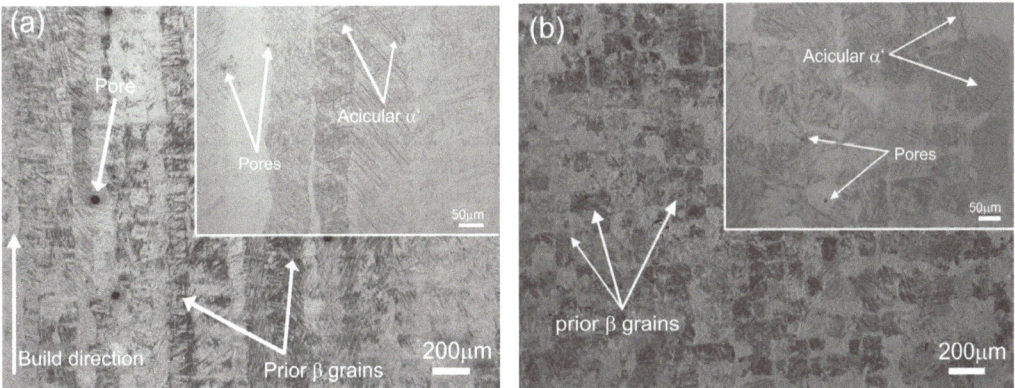

Fig. (7). Optical microstructure of SLM-produced Ti-6Al-4V sample: (**a**) XZ-plane and (**b**) XY-plane. The insets show the corresponding enlarged images [68].

Fig. (7) shows the optical microstructure of XY-plane and XZ-plane (build direction) for SLM-produced Ti-6Al-4V. It is clear to see that there still exist some differences in structural features between XZ-plane and XY-plane although both two planes are dominated by a large amount of acicular α′ martensite. Unlike the columnar prior β grains in XZ-plane (Fig. **7a**), the preceding β grains are square like in XY-plane (Fig. **7b**), which is closely related to the scan strategy applied in the SLM process. By quantitative examination the proportions of the distinct phases in the microstructure of SLM-created Ti-6Al-4V composite, it was discovered that the proportion of α′ phase in XY-plane is 88.1%, which means the β phase only contain 11.9%. However, the content of α′ phase and β phase are 95.0% and 5.0% in respectively the XZ-plane. The difference in this phase content is also reflected in the study of corrosion behavior. Because the proportion of α′-Ti in the XY plane is less than XZ plane, which directly leads to the inferior corrosion resistance of XZ-plane relative to XY-plane [68]. Furthermore, there is a difference in residual stress between the two planes of the SLM-produced Ti-6Al-4V. The residual stress tests in Ref [68] indicate that the estimation of residual stress for XY-plane is 115 MPa, which is slightly lower than that for XZ-plane (129 MPa). The quality of these two different planes seems to be very close.

Also, as mentioned before, the acicular α′ phase formed by high cooling rate will improve mechanical properties, so the large amount of α′ martensite formed in the make SLM-produced Ti-6Al-4V parts exhibit a higher the microhardness (409 Hv) than superplastic forming processed counterparts (346 Hv) [72]. Furthermore, SLM-produced Ti-6Al-4V samples have shown higher yield strength and higher ultimate tensile strength than the cast counterparts, resulting from the martensitic microstructure (dominant α′ martensite) formed during SLM process [34].

In addition, Ti-6Al-7Nb alloy is another (α+β)-type titanium alloy to replace Ti-6Al-4V; it has also been studied through SLM [73, 74]. The SLM-produced Ti-6Al-7Nb displays a similar microstructure to Ti-6Al-4V, which is composed of dominant α′ martensitic plates [73]. Depending on the final microstructural characteristics resulting from the SLM process, the SLM-produced Ti-6Al-7Nb exhibits Vickers microhardness ranging between 357 and 464 Hv [73]. Similarly, the tensile strength of the Ti-6Al-7Nb samples produced by SLM is primarily determined by the manufacturing strategies employed in the SLM process, which may result in residual stresses, structural defects, a number of holes and build layers arranged in the direction of tension. Similar to SLM-produced Ti-6Al-4V parts, SLM-produced Ti-6Al-7Nb samples also exhibit enhanced strength than its wrought samples and thermomechanically processed counterparts [72, 73]. The yield strength and ultimate tensile strength of SLM produced samples reach at 1440 MPa and 1515 MPa, respectively.

SLM of β-type Titanium Alloy

Compared to the wide range of studies on SLM produced Ti-6Al-4V, there is a limited research on the SLM-produced β-type titanium alloys due to the feedstock limitation of β-type titanium powder [30]. So far, only some β-type titanium alloys, such as Ti-24Nb-4Zr-8Sn (Ti2448), Ti-21Nb-17Zr, Ti-5553, Ti-50Ta) have been studied [35, 75 - 77]. The Ti2448 alloy is a β-type titanium alloy with low modulus and high strength combination in biomedical applications. Some studies have been conducted on the SLM produced β-type alloy with low-modulus and containing non-toxic elements [35, 52, 78]. As seen in Fig. (**8**), the Vickers microhardness and relative density of SLM-produced Ti2448 samples are strongly identified with the handling parameters in the SLM. In this figure, the relative density of the Ti2448 sample decreases with increasing scan speed and reaches a steady state at the scanning speed of 350 mm/s and 600 mm/s, where the relative density is exceeding 99%. To produce fully dense Ti2448 *via* SLM, the critical laser energy density is around 40 J/mm^3 [52], about one-third of 120 J/mm^3 [53, 54, 56] for manufacturing CP-Ti and Ti-6Al-4V. The microhardness achieves 220 Hv for the near completely dense parts. The modulus and elongation of all SLM-produced Ti2448 samples are closely tantamount to those for the counterpart processed by hot rolling [79] and by hot forging [80].

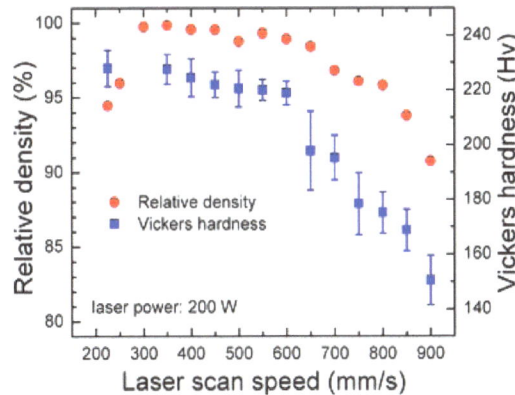

Fig. (8). Relative density and Vickers hardness of the SLM-produced Ti2448 at various laser scan speeds under the laser power of 200W [35].

Another near-β titanium alloy Ti-5553 can also be produced by SLM. Ti-5553 is known as a kind of attractive material with high variability, a wide range of mechanical properties and weldability, which can replace the applications of Ti-10-2-3 (Ti-10V-2Fe-3Al) in the aerospace industry. A relative density of 99.95% and a pure β-phase microstructure were obtained in the SLM-produced Ti-5553 alloy, and the obtained sample exhibited a tensile strength of about 800 MPa and strained up to 14%, as shown in Fig. (9), with a frail texturing along the building pivot on the (001) direction [77]. Furthermore, the shape of the grains of the Ti-5553 composites created by the SLM is extended with the morphology of a cellular to cellular-dendritic.

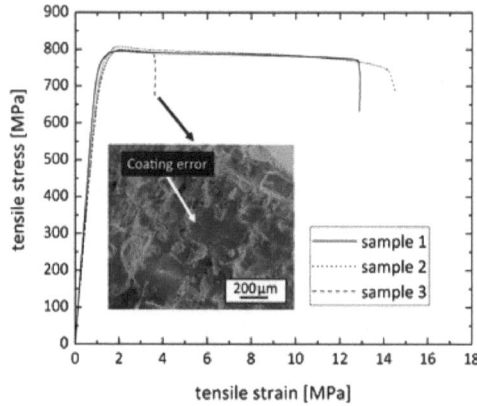

Fig. (9). Engineering stress-strain curves of SLM-produced Ti-5553 samples (inset shows the microstructure of the crack surface for sample 3) [77].

SLM OF POROUS TITANIUM STRUCTURES

SLM can not only build solid metal materials but also can produce lattice/porous structures [36, 58, 62, 75, 78, 81 - 85]. Metals with porous structure have been broadly applied to different ventures, such as load-bearing, heat exchange, impulse protection and damping [86]. The SLM innovations can provide the composites with specific shapes, and create directly from various engineering material powder (such as stainless steel, Ti alloys and Co-Cr alloys). SLM can also allow engineers to maximize freely customization and to use the porous structural materials with slightly losing its mechanical properties.

SLM produced porous titanium alloys like CP-Ti, Ti-6Al-4V, Ti-Nb-Zr and Ti-24Nb-4Zr-8Sn and their composites have been widely studied in order to meet different requirements, such as using the 55% to 75% porosity of the porous titanium structure alloy to simulate the human cancellous bone, [87]. The compressive strength of SLM-produced samples varies in the vicinity between 35 and 120 MPa. Similarly, Refs [84, 85, 88] also described some new design and manufacture of porous titanium alloy in its structures for improving bone development. It has additionally been exhibited that SLM is capable of producing advanced structures for bone growth and orthopedic devices. Moreover, the porous structure materials of CP-Ti and Ti-TiB with porosities of 10%, 17% and 37% have been efficiently fabricating by SLM [58]. Also the produced CP-Ti and

Ti-TiB composites have a closely mechanical performance to human bones, which includes yield strength (113-350 MPa for CP-Ti and 234-767 MPa for Ti-TiB) and elastic modulus (13-68 GPa for CP-Ti and 25-84 GPa for Ti-TiB composite). These parameters are demonstrating that they could be considered as a potential contender for biomedical implants.

Liu *et al.* [48] have examined the impact of various scanning parameters (5 different laser scan speeds ranging from 500 mm/s to 1500 mm/s) on the quality and mechanical properties of SLM produced biomedical Ti2448 alloy scaffolds. The relatively ideal manufacturing parameters were decided after analyzing the pores appropriation, geometrical precision and the mechanical properties (Fig. **10a** and **b**) [48]. When the laser power is set to 175 W and the scan speed changes between 750-1000 mm/s, the solid strut with the density above 99% for the scaffolds parts can be produced. The strength of the scaffold reaches 51 MPa in the use of these optimal processing parameters. The struts of these parts also have a very high density (>99%). However, by analyzing the fracture surface, the main reason caused the strut failure is the shortcoming of struts, such as the uniformed thickness, pores and unmelted powder inside substantial strut parts. The failure seems to happen in the flat arms which may convey the tensile load (Fig. **10c** and **d**) [48].

Liu *et al.* [47] also investigated the microstructure, defects and mechanical properties of SLM-produced Ti2448 porous structures. In the porous structure of the sample is composed of a number of octahedral structure, due to the formation of the small recoil pressure and high surface tension in SLM processing, the cone-shape defects caused by the molten pool are formed from inside the keyhole. As shown in Fig. (**11**) [47], the fatigue strength of the annealed SLM sample has a significant effect on the magnitude of the stress. At lower stress levels, the fatigue behavior of the porous material is dominated by the cyclic ratchet and surface properties. However, at higher stress levels, the fatigue life of the sample becomes very low and variable due to the generation and diffusion of cracks.

The porous structure alloys made by SLM are also examined and discussed on Ti-6Al-4V alloys. For example, SLM produced biomedical Ti-6Al-4V scaffolds alloy [83] has a very good biocompatibility and compressive capacity, which has been used in the culture of human osteoblasts and can be considered to have the potential for bone tissue engineering. Moreover, as detailed in Ref [12], SLM has the functional features of reproducing the complex microstructure. Ti-6Al-4V alloy has been successfully made of structural-cage shape. The average compressive modulus is 2.97 GPa; this value is desired and between trabecular (0.1-0.5 GPa) and cortical bone (15 GPa). Furthermore, theoretical and experimental measurements were conducted on octahedral Ti-6Al-4V porous structures in terms of compression test. The relationship between porosity porosity of octahedral porous structures in Ti-6Al-4V and experimental fracture load is exponential [89]. The failure of porous structures is brittle *via* cleavage fracture.

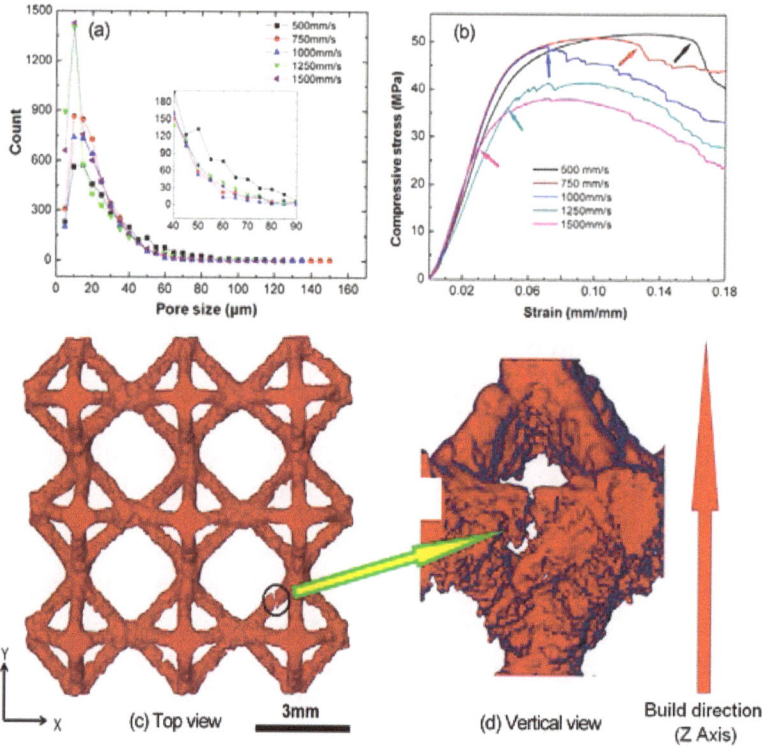

Fig. (10). Mechanical properties and geometric structure of SLM-processed Ti2448 scaffolds: (**a**) pores size with different laser scan speeds, (**b**) compressive stress-strain curves (the arrows indicating the location of the first strut failure, (**c**) the location of the failure (black circle) and (**d**) crack enlarged [48].

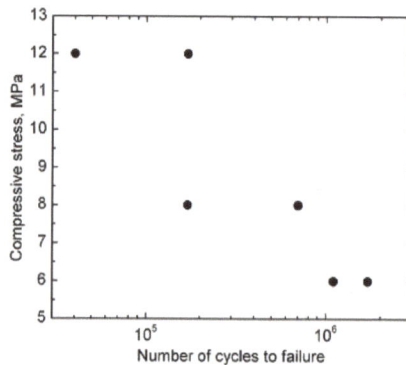

Fig. (11). The fatigue properties of the SLM-produced Ti-24Nb-4Zr-8Sn samples after annealing treatment at 750 °C for 1 hour [47].

CONCLUDING REMARKS

This chapter reviews a number of recent progress in the different types of titanium alloys (α-, α+β-, and β-type alloys) and their composites manufactured by selective laser melting (SLM). SLM could be used to produce titanium alloys with near full density (exceeding 99.5%) by optimal processing parameters. The SLM-produced different type of titanium alloys generally show finer microstructure and better mechanical properties than their counterparts processed by traditional technologies. Yet, more endeavors are needed for investigating the service properties, such as corrosion behavior and fatigue properties.

CONFLICT OF INTEREST

The author (editor) declares no conflict of interest, financial or otherwise.

ACKNOWLEDGEMENTS

This research was supported under the Australian Research Council's Discovery Projects (DP110101653). The authors are grateful to Y.J. Liu, H. Attar, J. Eckert, T.B Sercombe, Y.L. Hao, M. Calin, D. Klemm, K.G. Prashanth, M. Bönisch, S. Scudino, L. Löber, and A. Funk for collaborations.

REFERENCES

[1]   Banerjee D, Williams J. Perspectives on titanium science and technology. Prog Mater Sci 2013; 61: 844-79.

[2]   Geetha M, Singh AK, Asokamani R, Gogia AK. Ti based biomaterials, the ultimate choice for orthopaedic implants–a review. Prog Mater Sci 2009; 54: 397-425.
[http://dx.doi.org/10.1016/j.pmatsci.2008.06.004]

[3]   Liu LH, Yang C, Wang F, *et al.* Ultrafine grained Ti-based composites with ultrahigh strength and ductility achieved by equiaxing microstructure. Mater Des 2015; 79: 1-5.
[http://dx.doi.org/10.1016/j.matdes.2015.04.032]

[4]   Niinomi M. Fatigue performance and cyto-toxicity of low rigidity titanium alloy, Ti-29Nb-13-a-4.6Zr. Biomaterials 2003; 24(16): 2673-83.
[http://dx.doi.org/10.1016/S0142-9612(03)00069-3] [PMID: 12711513]

[5]   Calin M, Zhang LC, Eckert J. Tailoring of microstructure and mechanical properties of a Ti-based bulk metallic glass-forming alloy. Scr Mater 2007; 57: 1101-4.
[http://dx.doi.org/10.1016/j.scriptamat.2007.08.018]

[6]   Carman A, Zhang LC, Ivasishin OM, Savvakin DG, Matviychuk MV, Pereloma EV. Role of alloying elements in microstructure evolution and alloying elements behaviour during sintering of a near-[beta] titanium alloy. Mater Sci Eng A 2011; 528: 1686-93.
[http://dx.doi.org/10.1016/j.msea.2010.11.004]

[7]   Ehtemam-Haghighi S, Cao G, Zhang LC. Nanoindentation study of mechanical properties of Ti based alloys with Fe and Ta additions. J Alloys Compd 2017; 692: 892-7.
[http://dx.doi.org/10.1016/j.jallcom.2016.09.123]

[8]   Ehtemam-Haghighi S, Liu Y, Cao G, Zhang L-C. Phase transition, microstructural evolution and mechanical properties of Ti-Nb-Fe alloys induced by Fe addition. Mater Des 2016; 97: 279-86.

[http://dx.doi.org/10.1016/j.matdes.2016.02.094]

[9]   Ehtemam-Haghighi S, Liu Y, Cao G, Zhang LC. Influence of Nb on the β→α″ martensitic phase transformation and properties of the newly designed Ti-Fe-Nb alloys. Mater Sci Eng C 2016; 60: 503-10.
[http://dx.doi.org/10.1016/j.msec.2015.11.072] [PMID: 26706557]

[10]   Ehtemam-Haghighi S, Prashanth KG, Attar H, Chaubey AK, Cao GH, Zhang LC. Evaluation of mechanical and wear properties of TixNb7Fe alloys designed for biomedical applications. Mater Des 2016; 111: 592-9.
[http://dx.doi.org/10.1016/j.matdes.2016.09.029]

[11]   Haghighi Ehtemam S, Lu HB, Jian GY, Cao GH, Habibi D, Zhang LC. Effect of α″ martensite on the microstructure and mechanical properties of beta-type Ti-Fe-Ta alloys. Mater Des 2015; 76: 47-54.
[http://dx.doi.org/10.1016/j.matdes.2015.03.028]

[12]   Zhang LC, Das J, Lu HB, Duhamel C, Calin M, Eckert J. High strength Ti-Fe-Sn ultrafine composites with large plasticity. Scr Mater 2007; 57: 101-4.
[http://dx.doi.org/10.1016/j.scriptamat.2007.03.031]

[13]   Zhang LC, Kim KB, Yu P, Zhang WY, Kunz U, Eckert J. Amorphization in mechanically alloyed (Ti, Zr, Nb)-(Cu, Ni)-Al equiatomic alloys. J Alloys Compd 2007; 428: 157-63.
[http://dx.doi.org/10.1016/j.jallcom.2006.03.092]

[14]   Zhang LC, Lu HB, Mickel C, Eckert J. Ductile ultrafine-grained Ti-based alloys with high yield strength. Appl Phys Lett 2007; 91: 051906.
[http://dx.doi.org/10.1063/1.2766861]

[15]   Zhang LC, Shen ZQ, Xu J. Glass formation in a (Ti,Zr,Hf)-(Cu,Ni,Ag)-Al high-order alloy system by mechanical alloying. J Mater Res 2003; 18: 2141-9.
[http://dx.doi.org/10.1557/JMR.2003.0300]

[16]   Zhang LC, Xu J. Glass-forming ability of melt-spun multicomponent (Ti, Zr, Hf)-(Cu, Ni, Co)-Al alloys with equiatomic substitution. J Non-Cryst Solids 2004; 347: 166-72.
[http://dx.doi.org/10.1016/j.jnoncrysol.2004.09.007]

[17]   Zhang LC, Xu J, Ma E. Consolidation and properties of ball-milled Ti50Cu18Ni22Al4Sn6 glassy alloy by equal channel angular extrusion. Mater Sci Eng A 2006; 434: 280-8.
[http://dx.doi.org/10.1016/j.msea.2006.06.085]

[18]   Liu LH, Yang C, Kang LM, *et al.* Equiaxed Ti-based composites with high strength and large plasticity prepared by sintering and crystallizing amorphous powder. Mater Sci Eng A 2016; 650: 171-82.
[http://dx.doi.org/10.1016/j.msea.2015.10.048]

[19]   Blackman R, Barghi N, Tran C. Dimensional changes in casting titanium removable partial denture frameworks. J Prosthet Dent 1991; 65(2): 309-15.
[http://dx.doi.org/10.1016/0022-3913(91)90181-U] [PMID: 2051371]

[20]   Taira M, Moser JB, Greener EH. Studies of Ti alloys for dental castings. Dent Mater 1989; 5(1): 45-50.
[http://dx.doi.org/10.1016/0109-5641(89)90093-6] [PMID: 2691297]

[21]   Zhang LC, Das J, Lu HB, Duhamel C, Calin M, Eckert J. High strength Ti–Fe–Sn ultrafine composites with large plasticity. Scr Mater 2007; 57: 101-4.
[http://dx.doi.org/10.1016/j.scriptamat.2007.03.031]

[22]   Zhang LC, Xu J. Glass-forming ability of melt-spun multicomponent (Ti, Zr, Hf)–(Cu, Ni, Co)–Al alloys with equiatomic substitution. J Non-Cryst Solids 2004; 347: 166-72.
[http://dx.doi.org/10.1016/j.jnoncrysol.2004.09.007]

[23]   Yang Y, Li GP, Wang H, *et al.* Formation of zigzag-shaped {112}< 111 > beta mechanical twins in Ti-24.5 Nb-0.7 Ta-2 Zr-1.4 O alloy. Scr Mater 2012; 66: 211-4.

[http://dx.doi.org/10.1016/j.scriptamat.2011.10.031]

[24]   Lei X, Dong L, Zhang Z, *et al*. Microstructure, texture evolution and mechanical properties of VT3-1 titanium alloy processed by multi-pass drawing and subsequent isothermal annealing. Metals (Basel) 2017; 7: 131.
[http://dx.doi.org/10.3390/met7040131]

[25]   Ryan GE, Pandit AS, Apatsidis DP. Porous titanium scaffolds fabricated using a rapid prototyping and powder metallurgy technique. Biomaterials 2008; 29(27): 3625-35.
[http://dx.doi.org/10.1016/j.biomaterials.2008.05.032] [PMID: 18556060]

[26]   Ning CQ, Zhou Y. *In vitro* bioactivity of a biocomposite fabricated from HA and Ti powders by powder metallurgy method. Biomaterials 2002; 23(14): 2909-15.
[http://dx.doi.org/10.1016/S0142-9612(01)00419-7] [PMID: 12069332]

[27]   Carman A, Zhang LC, Ivasishin OM, Savvakin DG, Matviychuk MV, Pereloma EV. Role of alloying elements in microstructure evolution and alloying elements behaviour during sintering of a near-β titanium alloy. Mater Sci Eng A 2011; 528: 1686-93.
[http://dx.doi.org/10.1016/j.msea.2010.11.004]

[28]   Kwok PJ, Oppenheimer SM, Dunand DC. Porous Titanium by Electro-chemical Dissolution of Steel Space-holders. Adv Eng Mater 2008; 10: 820-5.
[http://dx.doi.org/10.1002/adem.200800072]

[29]   Chen YJ, Feng B, Zhu YP, Weng J, Wang JX, Lu X. Fabrication of porous titanium implants with biomechanical compatibility. Mater Lett 2009; 63: 2659-61.
[http://dx.doi.org/10.1016/j.matlet.2009.09.029]

[30]   Zhang LC, Attar H. Selective Laser Melting of Titanium Alloys and Titanium Matrix Composites for Biomedical Applications: A Review. Adv Eng Mater 2016; 18: 463-75.
[http://dx.doi.org/10.1002/adem.201500419]

[31]   Zhang LC, Attar H, Calin M, Eckert J. Review on manufacture by selective laser melting and properties of titanium based materials for biomedical applications. Mater Technol 2016; 31: 66-76.
[http://dx.doi.org/10.1179/1753555715Y.0000000076]

[32]   Prashanth KG, Shahabi HS, Attar H, *et al*. Production of high strength Al 85 Nd 8 Ni 5 Co 2 alloy by selective laser melting. Add Manuf 2015; 6: 1-5.

[33]   Prashanth KG, Scudino S, Klauss HJ, *et al*. Microstructure and mechanical properties of Al–12Si produced by selective laser melting: Effect of heat treatment. Mater Sci Eng A 2014; 590: 153-60.
[http://dx.doi.org/10.1016/j.msea.2013.10.023]

[34]   Thijs L, Verhaeghe F, Craeghs T, Van Humbeeck J, Kruth JP. A study of the microstructural evolution during selective laser melting of Ti–6Al–4V. Acta Mater 2010; 58: 3303-12.
[http://dx.doi.org/10.1016/j.actamat.2010.02.004]

[35]   Zhang LC, Klemm D, Eckert J, Hao YL, Sercombe TB. Manufacture by selective laser melting and mechanical behavior of a biomedical Ti-24Nb-4Zr-8Sn alloy. Scr Mater 2011; 65: 21-4.
[http://dx.doi.org/10.1016/j.scriptamat.2011.03.024]

[36]   Gu DD, Hagedorn YC, Meiners W, *et al*. Densification behavior, microstructure evolution, and wear performance of selective laser melting processed commercially pure titanium. Acta Mater 2012; 60: 3849-60.
[http://dx.doi.org/10.1016/j.actamat.2012.04.006]

[37]   Louvis E, Fox P, Sutcliffe CJ. Selective laser melting of aluminium components. J Mater Proc Tech 2011; 211: 275-84.
[http://dx.doi.org/10.1016/j.jmatprotec.2010.09.019]

[38]   Wang XJ, Zhang LC, Fang MH, Sercombe TB. The effect of atmosphere on the structure and properties of a selective laser melted Al–12Si alloy. Mater Sci Eng A 2014; 597: 370-5.
[http://dx.doi.org/10.1016/j.msea.2014.01.012]

[39]   Li XP, Wang XJ, Saunders M, *et al.* A selective laser melting and solution heat treatment refined Al–12Si alloy with a controllable ultrafine eutectic microstructure and 25% tensile ductility. Acta Mater 2015; 95: 74-82.
[http://dx.doi.org/10.1016/j.actamat.2015.05.017]

[40]   Li XP, Kang CW, Huang H, Zhang LC, Sercombe TB. Selective laser melting of an $Al_{86}Ni_6Y_{4.5}Co_2La_{1.5}$ metallic glass: Processing, microstructure evolution and mechanical properties. Mater Sci Eng A 2014; 606: 370-9.
[http://dx.doi.org/10.1016/j.msea.2014.03.097]

[41]   Badrossamay M, Childs TH. Further studies in selective laser melting of stainless and tool steel powders. Int J Mach Tools Manuf 2007; 47: 779-84.
[http://dx.doi.org/10.1016/j.ijmachtools.2006.09.013]

[42]   Li RD, Shi YS, Wang ZG, Wang L, Liu JH, Jiang W. Densification behavior of gas and water atomized 316L stainless steel powder during selective laser melting. Appl Surf Sci 2010; 256: 4350-6.
[http://dx.doi.org/10.1016/j.apsusc.2010.02.030]

[43]   Jia QB, Gu DD. Selective laser melting additive manufacturing of Inconel 718 superalloy parts: Densification, microstructure and properties. J Alloys Compd 2014; 585: 713-21.
[http://dx.doi.org/10.1016/j.jallcom.2013.09.171]

[44]   Amato KN, Gaytan SM, Murr LE, *et al.* Microstructures and mechanical behavior of Inconel 718 fabricated by selective laser melting. Acta Mater 2012; 60: 2229-39.
[http://dx.doi.org/10.1016/j.actamat.2011.12.032]

[45]   Scudino S, Unterdörfer C, Prashanth KG, *et al.* Additive manufacturing of CU-10Sn bronze. Mater Lett 2015; 156: 202-4.
[http://dx.doi.org/10.1016/j.matlet.2015.05.076]

[46]   Liu Y, Li S, Hou W, *et al.* Electron Beam Melted Beta-type Ti-24Nb-4Zr-8Sn Porous Structures With High Strength-to-Modulus Ratio. J Mater Sci Technol 2016; 32: 505-8.
[http://dx.doi.org/10.1016/j.jmst.2016.03.020]

[47]   Liu YJ, Li SJ, Wang HL, *et al.* Microstructure, defects and mechanical behavior of beta-type titanium porous structures manufactured by electron beam melting and selective laser melting. Acta Mater 2016; 113: 56-67.
[http://dx.doi.org/10.1016/j.actamat.2016.04.029]

[48]   Liu YJ, Li XP, Zhang LC, Sercombe TB. Processing and properties of topologically optimised biomedical Ti–24Nb–4Zr–8Sn scaffolds manufactured by selective laser melting. Mater Sci Eng A 2015; 642: 268-78.
[http://dx.doi.org/10.1016/j.msea.2015.06.088]

[49]   Liu YJ, Wang HL, Li SJ, *et al.* Compressive and fatigue behavior of beta-type titanium porous structures fabricated by electron beam melting. Acta Mater 2017; 126: 58-66.
[http://dx.doi.org/10.1016/j.actamat.2016.12.052]

[50]   Kruth JP, Mercelis P, Van Vaerenbergh J, Froyen L, Rombouts M. Binding mechanisms in selective laser sintering and selective laser melting. Rapid Prototyping J 2005; 11: 26-36.
[http://dx.doi.org/10.1108/13552540510573365]

[51]   Rigo O, Engel C. Selective Laser Melting versus Electron Beam Melting 2013. http://www.slideshare.net/carstenengel/selective-laser-melting-versus-electron-beam-melting

[52]   Zhang LC, Sercombe TB. Selective Laser Melting of Low-Modulus Biomedical Ti-24Nb-4Zr-8Sn Alloy: Effect of Laser Point Distance. Key Eng Mater 2012; 520: 226-33.
[http://dx.doi.org/10.4028/www.scientific.net/KEM.520.226]

[53]   Attar H, Bönisch M, Calin M, Zhang LC, Scudino S, Eckert J. Selective laser melting of *in situ* titanium–titanium boride composites: Processing, microstructure and mechanical properties. Acta Mater 2014; 76: 13-22.

[http://dx.doi.org/10.1016/j.actamat.2014.05.022]

[54]   Attar H, Calin M, Zhang LC, Scudino S, Eckert J. Manufacture by selective laser melting and mechanical behavior of commercially pure titanium. Mater Sci Eng A 2014; 593: 170-7.
[http://dx.doi.org/10.1016/j.msea.2013.11.038]

[55]   Aboulkhair NT, Everitt NM, Ashcroft I, Tuck C. Reducing porosity in AlSi10Mg parts processed by selective laser melting. Add Manuf 2014; 1: 77-86.

[56]   Gu DD, Hagedorn YC, Meiners W, Wissenbach K, Poprawe R. Nanocrystalline TiC reinforced Ti matrix bulk-form nanocomposites by Selective Laser Melting (SLM): Densification, growth mechanism and wear behavior. Comp Sci Tech 2011; 71: 1612-20.
[http://dx.doi.org/10.1016/j.compscitech.2011.07.010]

[57]   Barbas A, Bonnet AS, Lipinski P, Pesci R, Dubois G. Development and mechanical characterization of porous titanium bone substitutes. J Mech Behav Biomed Mater 2012; 9: 34-44.
[http://dx.doi.org/10.1016/j.jmbbm.2012.01.008] [PMID: 22498281]

[58]   Attar H, Löber L, Funk A, et al. Mechanical behavior of porous commercially pure Ti and Ti–TiB composite materials manufactured by selective laser melting. Mater Sci Eng A 2015; 625: 350-6.
[http://dx.doi.org/10.1016/j.msea.2014.12.036]

[59]   Lin CW, Ju CP, Lin JH. Comparison among mechanical properties of investment-cast cp Ti, Ti-6A--7Nb and Ti-15Mo-1Bi alloys. Mater Trans 2004; 45: 3028-32.
[http://dx.doi.org/10.2320/matertrans.45.3028]

[60]   Güçlü FM, Çimenoğlu H. The Recrystallization Behaviour Of CP-Titanium. Mater Sci Forum 2004; 467-470: 459-64.
[http://dx.doi.org/10.4028/www.scientific.net/MSF.467-470.459]

[61]   Attar H, Prashanth KG, Chaubey AK, et al. Comparison of wear properties of commercially pure titanium prepared by selective laser melting and casting processes. Mater Lett 2015; 142: 38-41.
[http://dx.doi.org/10.1016/j.matlet.2014.11.156]

[62]   Attar H, Bönisch M, Calin M, et al. Comparative study of microstructures and mechanical properties of in situ Ti–TiB composites produced by selective laser melting, powder metallurgy, and casting technologies. J Mater Res 2014; 29: 1941-50.
[http://dx.doi.org/10.1557/jmr.2014.122]

[63]   Zhang LC, Xu J, Eckert J. Thermal stability and crystallization kinetics of mechanically alloyed TiC/Ti-based metallic glass matrix composite. J Appl Phys 2006; 100: 033514.
[http://dx.doi.org/10.1063/1.2234535]

[64]   Facchini L, Magalini E, Robotti P, Molinari A, Höges S, Wissenbach K. Ductility of a Ti-6Al-4V alloy produced by selective laser melting of prealloyed powders. Rapid Prototyping J 2010; 16: 450-9.
[http://dx.doi.org/10.1108/13552541011083371]

[65]   Sallica-Leva E, Jardini AL, Fogagnolo JB. Microstructure and mechanical behavior of porous Ti-6A--4V parts obtained by selective laser melting. J Mech Behav Biomed Mater 2013; 26: 98-108.
[http://dx.doi.org/10.1016/j.jmbbm.2013.05.011] [PMID: 23773976]

[66]   Dai N, Zhang J, Chen Y, Zhang LC. Heat Treatment Degrading the Corrosion Resistance of Selective Laser Melted Ti-6Al-4V Alloy. J Electrochem Soc 2017; 164: C428-34.
[http://dx.doi.org/10.1149/2.1481707jes]

[67]   Dai N, Zhang LC, Zhang J, Chen Q, Wu M. Corrosion behavior of selective laser melted Ti-6Al-4 V alloy in NaCl solution. Corros Sci 2016; 102: 484-9.
[http://dx.doi.org/10.1016/j.corsci.2015.10.041]

[68]   Dai N, Zhang LC, Zhang J, et al. Distinction in corrosion resistance of selective laser melted Ti-6A--4V alloy on different planes. Corros Sci 2016; 111: 703-10.
[http://dx.doi.org/10.1016/j.corsci.2016.06.009]

[69]   Thijs L, Verhaeghe F, Craeghs T, Van Humbeeck J, Kruth JP. A study of the microstructural evolution during selective laser melting of Ti-6Al-4V. Acta Mater 2010; 58: 3303-12.
[http://dx.doi.org/10.1016/j.actamat.2010.02.004]

[70]   Bai Y, Gai X, Li S, *et al.* Improved corrosion behaviour of electron beam melted Ti-6Al–4V alloy in phosphate buffered saline. Corros Sci 2017; 123: 289-96.
[http://dx.doi.org/10.1016/j.corsci.2017.05.003]

[71]   Das M, Balla VK, Basu D, Bose S, Bandyopadhyay A. Laser processing of SiC-particle-reinforced coating on titanium. Scr Mater 2010; 63: 438-41.
[http://dx.doi.org/10.1016/j.scriptamat.2010.04.044]

[72]   Niinomi M. Mechanical properties of biomedical titanium alloys. Mater Sci Eng A 1998; 243: 231-6.
[http://dx.doi.org/10.1016/S0921-5093(97)00806-X]

[73]   Chlebus E, Kuźnicka B, Kurzynowski T, Dybała B. Microstructure and mechanical behaviour of Ti-6Al-7Nb alloy produced by selective laser melting. Mater Charact 2011; 62: 488-95.
[http://dx.doi.org/10.1016/j.matchar.2011.03.006]

[74]   Marcu T, Todea M, Gligor I, Berce P, Popa C. Effect of surface conditioning on the flowability of Ti6Al7Nb powder for selective laser melting applications. Appl Surf Sci 2012; 258: 3276-82.
[http://dx.doi.org/10.1016/j.apsusc.2011.11.081]

[75]   Speirs M, Van Humbeeck J, Schrooten J, Luyten J, Kruth JP. The effect of pore geometry on the mechanical properties of selective laser melted Ti-13Nb-13Zr scaffolds. Procedia CIRP 2013; 5: 79-82.
[http://dx.doi.org/10.1016/j.procir.2013.01.016]

[76]   Sing SL, Yeong WY, Wiria FE. Selective laser melting of titanium alloy with 50 wt% tantalum: Microstructure and mechanical properties. J Alloys Compd 2016; 660: 461-70.
[http://dx.doi.org/10.1016/j.jallcom.2015.11.141]

[77]   Schwab H, Palmb F, Kühn U, Eckert J. Microstructure and mechanical properties of the near-beta titanium alloy Ti-5553 processed by selective laser melting. Mater Des 2016; 105: 75-80.
[http://dx.doi.org/10.1016/j.matdes.2016.04.103]

[78]   Liu YJ, Li XP, Zhang LC, Sercombe TB. Processing and properties of topologically optimised biomedical Ti-24 Nb-4Zr-8Sn scaffolds manufactured by selective laser melting. Mater Sci Eng A 2015; 642: 268-78.
[http://dx.doi.org/10.1016/j.msea.2015.06.088]

[79]   Zhang SQ, Li SJ, Jia MT, Hao YL, Yang R. Fatigue properties of a multifunctional titanium alloy exhibiting nonlinear elastic deformation behavior. Scr Mater 2009; 60: 733-6.
[http://dx.doi.org/10.1016/j.scriptamat.2009.01.007]

[80]   Li SJ, Cui TC, Hao YL, Yang R. Fatigue properties of a metastable β-type titanium alloy with reversible phase transformation. Acta Biomater 2008; 4(2): 305-17.
[http://dx.doi.org/10.1016/j.actbio.2007.09.009] [PMID: 18006397]

[81]   Vandenbroucke B, Kruth JP. Selective laser melting of biocompatible metals for rapid manufacturing of medical parts. Rapid Prototyping J 2007; 13: 196-203.
[http://dx.doi.org/10.1108/13552540710776142]

[82]   Yadroitsev I, Shishkovsky I, Bertrand P, Smurov I. Manufacturing of fine-structured 3D porous filter elements by selective laser melting. Appl Surf Sci 2009; 255: 5523-7.
[http://dx.doi.org/10.1016/j.apsusc.2008.07.154]

[83]   Warnke PH, Douglas T, Wollny P, *et al.* Rapid prototyping: porous titanium alloy scaffolds produced by selective laser melting for bone tissue engineering. Tissue Eng Part C Methods 2009; 15(2): 115-24.
[http://dx.doi.org/10.1089/ten.tec.2008.0288] [PMID: 19072196]

[84]   Challis VJ, Roberts AP, Grotowski JF, Zhang LC, Sercombe TB. Prototypes for bone implant scaffolds designed *via* topology optimization and manufactured by solid freeform fabrication. Adv Eng Mater 2010; 12: 1106-10.
[http://dx.doi.org/10.1002/adem.201000154]

[85]   Challis VJ, Xu XX, Zhang LC, Roberts AP, Grotowski JF, Sercombe TB. High specific strength and stiffness structures produced using selective laser melting. Mater Des 2014; 63: 783-8.
[http://dx.doi.org/10.1016/j.matdes.2014.05.064]

[86]   Zhang Y, Ha S, Sharp K, Guest JK, Weihs TP, Hemker KJ. Fabrication and mechanical characterization of 3D woven Cu lattice materials. Mater Des 2015; 85: 743-51.
[http://dx.doi.org/10.1016/j.matdes.2015.06.131]

[87]   Pattanayak DK, Fukuda A, Matsushita T, *et al.* Bioactive Ti metal analogous to human cancellous bone: Fabrication by selective laser melting and chemical treatments. Acta Biomater 2011; 7(3): 1398-406.
[http://dx.doi.org/10.1016/j.actbio.2010.09.034] [PMID: 20883832]

[88]   Mullen L, Stamp RC, Brooks WK, Jones E, Sutcliffe CJ. Selective Laser Melting: a regular unit cell approach for the manufacture of porous, titanium, bone in-growth constructs, suitable for orthopedic applications. J Biomed Mater Res B Appl Biomater 2009; 89(2): 325-34.
[http://dx.doi.org/10.1002/jbm.b.31219] [PMID: 18837456]

[89]   Sun JF, Yang YQ, Wang D. Mechanical properties of a Ti6Al4V porous structure produced by selective laser melting. Mater Des 2013; 49: 545-52.
[http://dx.doi.org/10.1016/j.matdes.2013.01.038]

CHAPTER 5

Electron Beam Melting of Porous Titanium Alloys: Microstructure and Mechanical Behavior

Lai-Chang Zhang[1,*], **Yujing Liu**[1] and **Liqiang Wang**[2]

[1] *School of Engineering, Edith Cowan University, Perth, WA, Australia*

[2] *State Key Laboratory of Metal Matrix Composites, Shanghai Jiao Tong University, Shanghai, China*

Abstract: Electron beam melting (EBM) is a relatively new rapid, additive manufacturing technology which is capable of fabricating complex, multi-functional metal or alloy components directly from CAD models, selective melting of precursor powder beds. Compared with Ti-6Al-4V samples with same porosity level, the EBM-produced β-type Ti-24Nb-4Zr-8Sn (Ti2448) porous components exhibit a higher normalized fatigue strength owing to super-elastic property, greater plastic zone ahead of the fatigue crack tip and the crack deflection behavior. The super-elastic property can be improved by increasing porosity of porous samples as a result of increasing the tensile/compressive stress ratio of the porous structure. EBM-produced components exhibit more than twice the strength-to-modulus ratio of porous Ti-6Al-4V counterparts. The position of fatigue crack initiation is defined in strain curves based on the variation of the fatigue cyclic loops. The unique manufacturing process of EBM results in the generation of different sizes of grains, and the apparent fatigue crack deflection occurs at the grain boundaries in the columnar grain zone due to substantial misorientation between adjacent grains.

Keywords: Electron beam melting, Mechanical properties, Microstructure, Porous material, Titanium alloys.

INTRODUCTION

Recently, the demand for implants is increasing as more people are suffering from joint problems caused by aging population and obesity [1, 2]. It is therefore necessary to produce high quality, artificial joints to reduce the risk of revision surgery. Several requirements, such as desirable customized complex shape to fit surrounding bone, interconnecting porosity with suitable size to facilitate bone in-growth, high strength and low Young's modulus of the material used, are needed

---

* **Corresponding author Lai-Chang Zhang:** School of Engineering, Edith Cowan University, 270 Joondalup Drive, Joondalup, Perth, WA 6027, Australia; Tel: 61 8 63042322; Fax: 61 8 63045811; E-mails: lczhangimr@gmail.com; l.zhang@ecu.edu.au

Liqiang Wang & Lai-Chang Zhang (Eds.)

to satisfy a successful implant [3]. Fortunately, additive manufacturing (AM) is emerging as an excellent type of manufacturing technologies that is capable of manufacturing porous implants with optimal properties to meet these requirements by using medical grade metallic powder materials [4]. The AM technologies can realize components by using layer-wise method from 3D models and from powder material, which have attracted increased interest in past decades.

Compared to conventional processing technologies, AM can make components with complicated geometries (such as porous structures) in shorter time and with less cost [5]. Usually, the as-fabricated components by AM technologies display a fine and different microstructure compared to those produced by conventional processing technologies, thereby exhibiting outstanding properties including low density, high strength, high toughness and large ductility [6 - 10]. Among the AM technologies, electron beam melting (EBM) is a common technology used to fabricate complex structures with high relative density and good mechanical properties. During EBM process, a heat source focuses on powder bed to scan the powder in a selected region. As such, the selected powder is melted then solidifies rapidly. Once the scan of one layer is completed, the build platform descends by one preset layer thickness and a new layer of powder is deposited on its top. Such a layer-by-layer process continues until the entire component is completely produced. As such, a real component is manufactured from its 3D model.

Thanks to low density, low Young's modulus, high strength, and high corrosion resistance and biocompatibility, titanium alloys are regarded as the most appropriate implant materials for load-bearing applications [11 - 14]. Currently, majority of the studies on AM-produced titanium alloys have been focused on the processing and mechanical properties of (α+β)-type Ti-6Al-4V. It has been reported that the AM-produced Ti-6Al-4V porous structures exhibit high biocompatibility, good mechanical properties and good corrosion resistance [15]. However, there is a concern that the toxic elements Al and V in Ti-6Al-4V might lead to allergic reaction and Alzheimer's disease [16]. Furthermore, the large mismatch in Young's modulus between Ti-6Al-4V implants and the surrounding bone can lead to the well-known "stress-shielding" phenomenon [17]. In addition, α' martensite usually forms in AM-produced Ti-6Al-4V components, which is detrimental to the ductility and fatigue life as well as to the corrosion resistance property [18]. Therefore, it is imperative to eliminate the above drawbacks.

β-type titanium alloys, such as Ti-29Nb-13Ta-4.6Zr, Ti-35Nb-5Ta-7Zr and Ti-24Nb-4Zr-8Sn (abbreviated as Ti2448), are attracting increasing research interest due to their advantages of low modulus and non-toxic elements [11, 19, 20]. For example, β-type Ti2448 has exhibited a lower modulus of ~42-50 GPa compared with (α+β)-type titanium alloys (~100-120 GPa), coupled with great biocompatibility and mechanical properties [9, 21 - 23]. Ti2448 has been successfully manufactured to dense and porous components by using AM methods [17, 24]. The AM-manufactured Ti2448 porous structure with designed porosity of 85% exhibits high relative density (~99.3%), low Young's modulus (~1 GPa) and high compressive strength (51 MPa) [8]. Ti2448 solid parts obtained *via* EBM at preheating temperature of ~200 °C consist of large columnar grains aligned with the build direction and possess high hardness (~2.5 GPa) [24], which is higher than that of AM-fabricated sample (~2.3 GPa) [17]. There is a lack of assessing how the microstructure of Ti2448 affects fatigue properties, which could be influenced by toughness, super-elastic properties, the crack-tip plastic zone size and crack deflection behaviors. Specifically, fatigue crack propagation behaviors may be affected by grain boundaries and the misorientation angle between adjacent grains. It was reported that dislocations or the slip-band might be blocked by grain boundaries [23, 24]. Thus the crack propagation may be deflected if the neighboring grains resist dislocations or the slip-band. Such a crack deflection behavior could play an active role in fatigue properties.

This chapter shows the influence of porosity variation on the mechanical properties of the β-type Ti2448 alloy porous samples, in terms of Young's modulus, super-elastic property, strength and fatigue properties. The relationship between the misorientation angle between adjacent grains and the fatigue crack deflection behaviors are also discussed.

MANUFACTURING PROCESS

Fig. (1) shows the single unit of 3D modelling and the EBM-produced porous sample morphology. The surface roughness, inside defects and microstructures are affected by the EBM manufacturing process. It is known that both EBM manufacturing processes are very complex and highly dynamic, where the thermal and temperature distribution of the melt pool has significant effect on the resulting microstructure. However, until now, there is no accurate method to measure the temperature of melting zone in EBM technology. Only the substrate temperature is measured by a sensor underneath the plate. From the temperature data collected from the plate, the powder bed temperature of processing area can be assumed. The preheating temperature of ~500 °C (note that this temperature is material-specific) in the EBM process. During the build phase, EBM samples were cooled relatively slowly due to the presence of a vacuum, and were kept between 400-500 °C for several hours. This is, in effect, identical to an aging treatment for Ti2448 alloy. This explains the presence of α phase in the as-fabricated EBM samples. The initial α grains nucleate at the β grain boundaries and then grow toward the center of β grains. Although most α grains are distributed along the β grain boundaries, a small fraction of α grains are located in the interior of β grains.

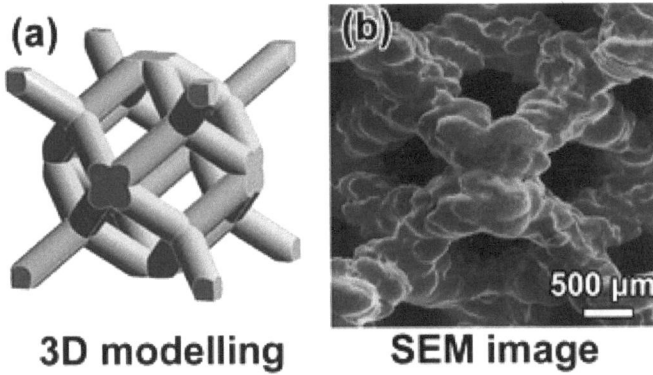

(a)

(b)

3D modelling   **SEM image**

Fig. (1). (**a**) The single unit of 3D rhombic dodecahedron modelling, (**b**) the morphology of the EBM-produced porous specimen.

Fig. (**2**) schematically displays the process of EBM. The input energy and momentum of the electron beam can directly affect the size of melt pool and the amount of defects. Following the suggestions by Simchi [25], the energy density (*Q*) can be affected by the processing parameters:

$$Q = \frac{\pi \eta P}{4dv} \qquad (1)$$

where P is the input power, η is the coupling efficiency, d is the diameter of electron/laser beam spot and v is the scan speed. According to Eq. (1), higher input energy will result in higher energy density, and thereby creating a larger melting zone. It is reported that the formation of a keyhole has a significant effect on the melt zone depth. Semak *et al.* [26] depicted that a recoil pressure forms as a result of metal vaporization and acts to push the liquid away from the melt zone, thereby generating a keyhole. The electron/laser beam can be reflected by the inner surface of the keyhole and the power is then concentrated on the bottom of keyhole (this process is shown in Fig. **2a** and **b**). This causes a higher temperature at the bottom than the top of the melting zone. Thereby, the electron beam can penetrate and melt a deep zone. The electron beam melting area of single scanning track is larger than 3D modelling design with a width of ~280±23 μm in x-y cross-section and a depth of ~152±15 μm in x-z cross-section (along building direction) (Fig. **2c**).

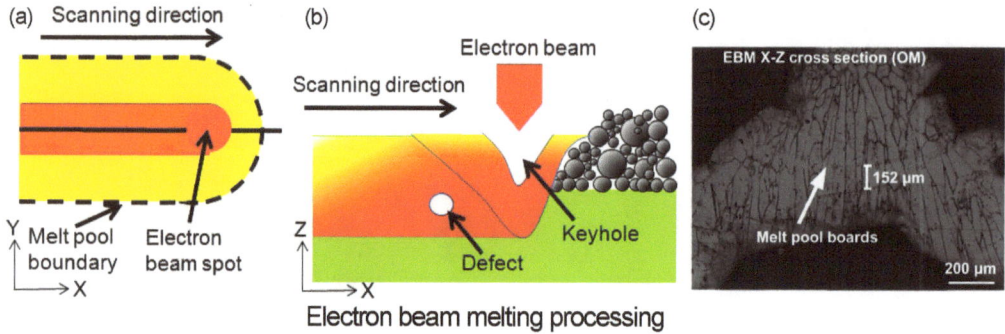

Fig. (2). The schematic diagram of the EBM melting process.

Defect and Surface Roughness Generation

The defects generated during EBM process will affect the performance of the as-fabricated samples. The mechanism of defects formation during additive manufacturing process is complex and not yet fully understood. For most metallic powder, the reported possible reasons for defects formation include material powder defects, insufficient energy, balling effects, material vaporization and imperfect collapse of the keyhole. Gong *et al*. [27] pointed out that the defects can have a negative influence on the tensile and fatigue properties of EBM-produced samples. Therefore, a detailed understanding is needed on mechanism of defect formation.

During the EBM build process, the heat source has approximate Gauss distribution and focuses on the powder bed [28, 29]. There are many physical effects occurring inside the melting zone including the interactions between solid, liquid and gas phases [30, 31], and the evaporation of the material which generates the recoil pressure and vapors capillary (named keyhole) [32, 33]. For Ti2448 alloy, the boiling point of each element is different. In particular tin, having the lowest boiling point of 2600 °C, would be the first to vaporize. Indeed, high concentrations of tin have been detected on the surface of pores by using EPMA (Fig. **3**), indicating that the vaporization of tin is playing a critical role in the formation of these pores.

MECHANICAL PROPERTIES OF POROUS STRUCTURES

In general, mechanical properties, such as the Young's modulus, compressive strength and fatigue properties, are influenced by the porosity of a structure. The hip and knee joints must tolerate cyclic loading when people walking, thus the improvement of mechanical attributes, especially fatigue properties, is required.

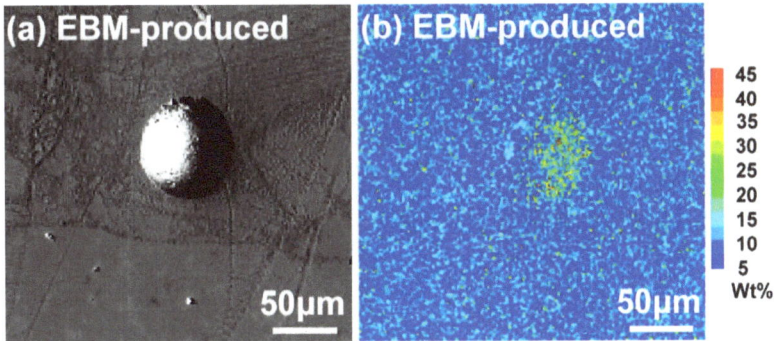

Fig. (3). (**a**) The SEM microstructure, and (**b**) EPMA quantitative chemical analysis maps for elemental tin near a defect for the EBM samples. Tin appears to concentrates on the surface of the defects.

Further understanding the mechanical properties of structures with different porosity is necessary to optimize porosity when designing implants. Ti-6Al-4V is a widely common applied titanium alloy and its porous structure behaviors, including microstructure and the mechanical properties, have been studied extensively [34 - 37]. The study on Ti-6Al-4V porous sample with different relative densities 0.73 to 1.68 g/cm^3 [37] has demonstrated that compressive strength and Young's modulus decrease with increasing porosity and that the interaction of the ratcheting effect and fatigue damage mechanism could adversely affect fatigue properties. Furthermore, the high Young's modulus and brittle deformation behavior, caused by the α' phase formation inside struts, may cause a "stress shielding" effect or reduce the life of the implant in humans. Importantly, recent studies revealed that the Ti-6Al-4V alloy would be harmful to patient's health as Al and V elements are released [38 - 40]. Therefore, non-toxic β-type titanium Ti2448 alloys with low modulus have been attracting increased attention. As mention above, the original EBM-produced samples consist of α+β phase. The porous rhombic dodecahedron structures containing 7×7×14 unit cells with nominal porosity of 67.9%, 72.5%, 75.0%, 77.4%, 79.5 and 91.2%, which were defined as A, B, C, D, E and F groups, were fabricated by EBM system. In order to study the properties of single β phase porous samples, the porous samples are annealed for 1 hour at 750°C to remove the α phase.

Super-elasticity

The loading-unloading curves for EBM-produced Ti2448 porous samples with different porosity are shown in Fig. (**4a**). All the sample groups were applied to cyclic uniaxial compressive loading with a total strain of 2-3% at a strain step of 0.5%. The results show that all the samples across the entire porosity range exhibit excellent super-elasticity property (only the samples with the porosity with 67.9%, 77.4% and 91.2% are displayed). Interestingly, the super-elastic property increases with increasing sample porosity. The sample with 91.2% porosity is almost fully recovered at 3% elastic strain, but the sample with 67.9% porosity exhibits weaker super-elastic property by only recovering to 0.5%. Such a difference in super-elastic property was explained by Finite Element Method (FEM) analysis. The super-elastic property can be affected by the ratio of tensile stress/compressive stress. The stress distributions of the models for 67.9% and 91.2% porosity samples, at the 3% compressive strain and the compressive stresses of 35.8 MPa (67.9%) and 7.2 MPa (91.2%), are plotted in Fig. (**4b**).

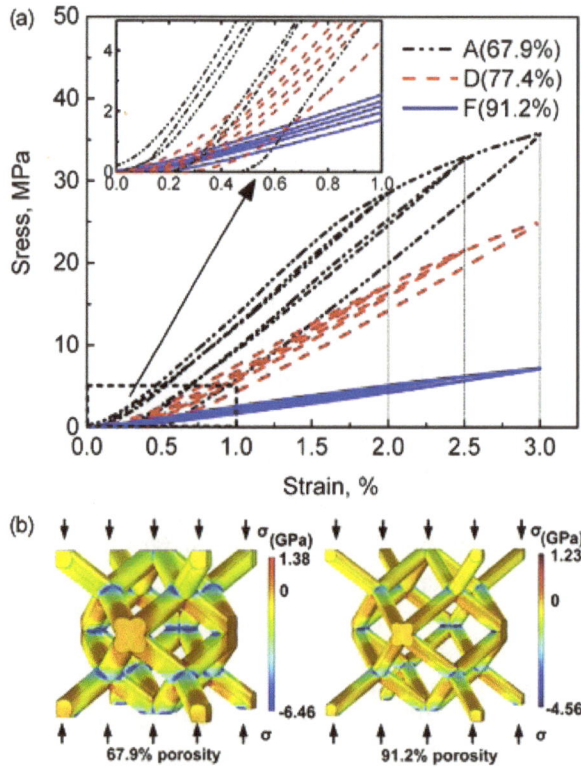

Fig. (4). (**a**) The variation of the superelastic property with different porosity, the superelastic property increases with the growing porosity, (**b**) FEM stress distribution of the rhombic dodecahedron structures with different porosity at the 3% compressive strain.

The tensile and compressive stresses are distributed on the top and bottom of each node, respectively. The ratio of maximum tensile stress/compressive stress for the samples with 67.9% and 91.2% porosity are 0.21 and 0.27, respectively. The tensile stress is the dominant factor for the super-elastic property [41]; the tensile/compressive stress ratio increases due to an increase in the porosity, and consequently indicates that a higher ratio of the tensile/compressive stress ratio enhances the super-elasticity property. Such a variation of tensile/compressive stress ratio can be considered as a function of increasing strain or samples with different porosity, at the same stress level, a better super-elasticity property would result in a greater strain.

The uniaxial compression curves for the Ti2448 samples with different porosity are shown in Fig. (**5a**). All the samples exhibit large plasticity without apparent layer-wise fracture. Clearly, both the compressive strength and yield strength decrease with increasing porosity. Compared with the Ti-6Al-4V sample with 75% porosity build from the same unit shape, which possesses a compressive strength of ~60 MPa [37], the Ti2448 sample shows a much lower compressive strength of ~38 MPa.

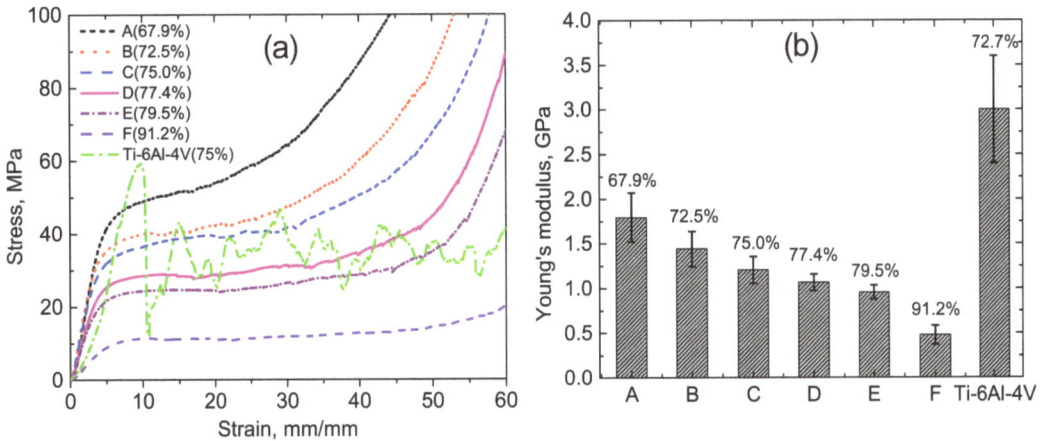

Fig. (5). The typical compressive stress-strain curves (**a**) and the Young's modulus (**b**) for Ti2448 annealed specimens and Ti-6Al-4V with different porosity.

In general, the relationship between the compressive stress and the relative density for the porous structure can be described by the Gibson-Ashby model [42]:

$$\sigma/\sigma_0 = C(\rho/\rho_0)^n \tag{2}$$

where the σ and ρ are the strength and density of fully dense samples, while σ_0 and ρ_0 are the strength and density of porous samples respectively. C is a constant, n is an exponential factor. Herein, the factor value of n is ~2.05 which is lower than that of Ti-6Al-4V (~2.7) [37]. Further, the samples with large porosity exhibit low Young's modulus (Fig. **5b**). The Ti2448 groups with single β phase show much lower Young's modulus than Ti-6Al-4V. For example, the Young's modulus of Ti2448 samples with a porosity of 72.5% is ~1.44±0.2 GPa, which is only half of the Ti-6Al-4V samples with 72.5% porosity (3.0±0.6 GPa).

Fatigue Properties

The total strain curves and the cyclic loops of the group B, in relation to high stress and low stress levels, are shown in Fig. (**6**). In general, the fatigue strain curves of the porous structures for both stress levels can be divided into three stages [43]. The first stage is only related to the plastic deformation accumulation, which is caused by ratcheting effect [36, 44]. The fatigue crack initiation or early strut damage is generated at the beginning of the second stage and the strain curves during this stage can be affected by the combination of plastic deformation and fatigue crack propagation. The third stage is the fatigue crack coalescence stage, in which the fatigue crack propagation rate increases dramatically and fatigue failure occurs. Some authors have divided the strain curve into three stages based on the slope of the static strain curve method [43, 45]. Unfortunately, it is difficult to determine the crack initiation position/early strut damage on the curves for the high stress level as the fatigue crack is initiated too early.

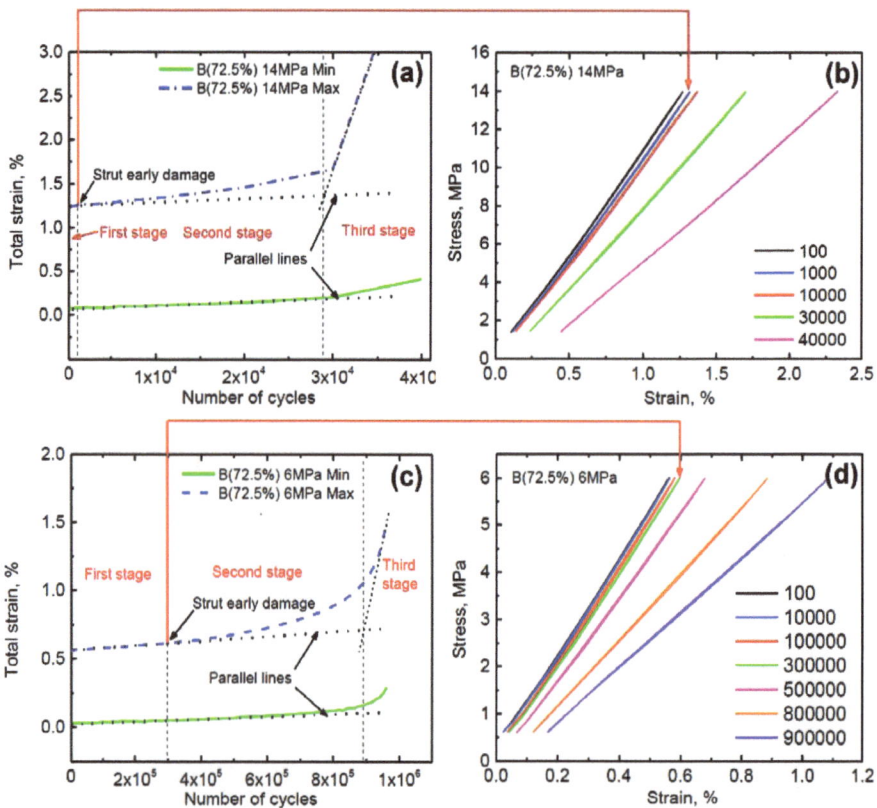

Fig. (6). The total fatigue strain of the annealed specimens with 72.5% porosity under the applied cyclic stress of (**a**) 14 MPa and (**c**) 6 MPa, the effect of fatigue cycle on the stress-strain loops of annealed specimens for (**b**) 14 MPa and (**d**) 6 MPa.

Here, we present a new method to define the crack initiation position (Fig. **6c**). Both the maximum and minimum strain curves need to be plotted in Fig. (**6**). Then a slope line is drawn for the minimum strain curve from the rising point of the strain curve. Another line is drawn parallel to the slope line and this line must pass through the rising point of the maximum strain curve. The position where the angle generation occurs between the line and the maximum strain curve is the crack initiation formation point. It can be seen that the crack initiation or strut early damage point corresponds to the cycle loops, where the dynamic Young's modulus begins to decrease (Fig. **6b** and **d**). Clearly, the dynamic Young's modulus decreases with the increasing cycles due to the fatigue crack occurring in the strut. It is evident that the crack initiation in the high stress level group occurring very early (Fig. **6a**) with the first stage is shorter than that of the low stress level (Fig. **6c**).

The strain accumulation for per cycle $d\varepsilon/dN$ (ratcheting rate) with different porosity is plotted in Fig. (**7a**). Both the applied stress and the porosity play dominant roles in influencing $d\varepsilon/dN$; a higher stress and higher porosity lead to a higher ratcheting rate $d\varepsilon/dN$. Interestingly, the excellent super-elastic property of the Ti2448 samples leads to a lower ratcheting rate compared with the Ti-6Al-4V samples with the same porosity and same structure [37].

The strain accumulation for per cycle $d\varepsilon/dN$ (ratcheting rate) with different porosity is plotted in Fig. (**7a**). Both the applied stress and the porosity play dominant roles in influencing $d\varepsilon/dN$; a higher stress and higher porosity lead to a higher ratcheting rate $d\varepsilon/dN$. Interestingly, the excellent super-elastic property of the Ti2448 samples leads to a lower ratcheting rate compared with the Ti-6Al-4V samples with the same porosity and same structure [37].

The S-N curves of all samples with absolute and normalized stress values are plotted in Fig. (**7b**). For Ti2448 samples, the fatigue life is dominated by the porosity and applied stress level; the fatigue life decreases with increasing porosity for both low and high stress levels, as reported in [37]. The high-porosity samples (such as F group with ~ 91.2% porosity) present much lower fatigue strength than low-porosity groups. However, they exhibit close normalized fatigue strength (Fig. **7c**), which is also much better than that for the Ti-6Al-4V samples with 75% porosity. Fig. (**7d**) shows the relationship between the fatigue strength and Young's modulus for the Ti2448 and Ti-6Al-4V samples. Clearly, the Young's modulus of Ti2448 samples is much lower than that of Ti-6Al-4V samples with the same fatigue strength. For example, at the fatigue strength of 4 MPa, the Young's modulus of Ti2448 and Ti-6Al-4V is ~1.44 GPa and ~3 GPa, respectively.

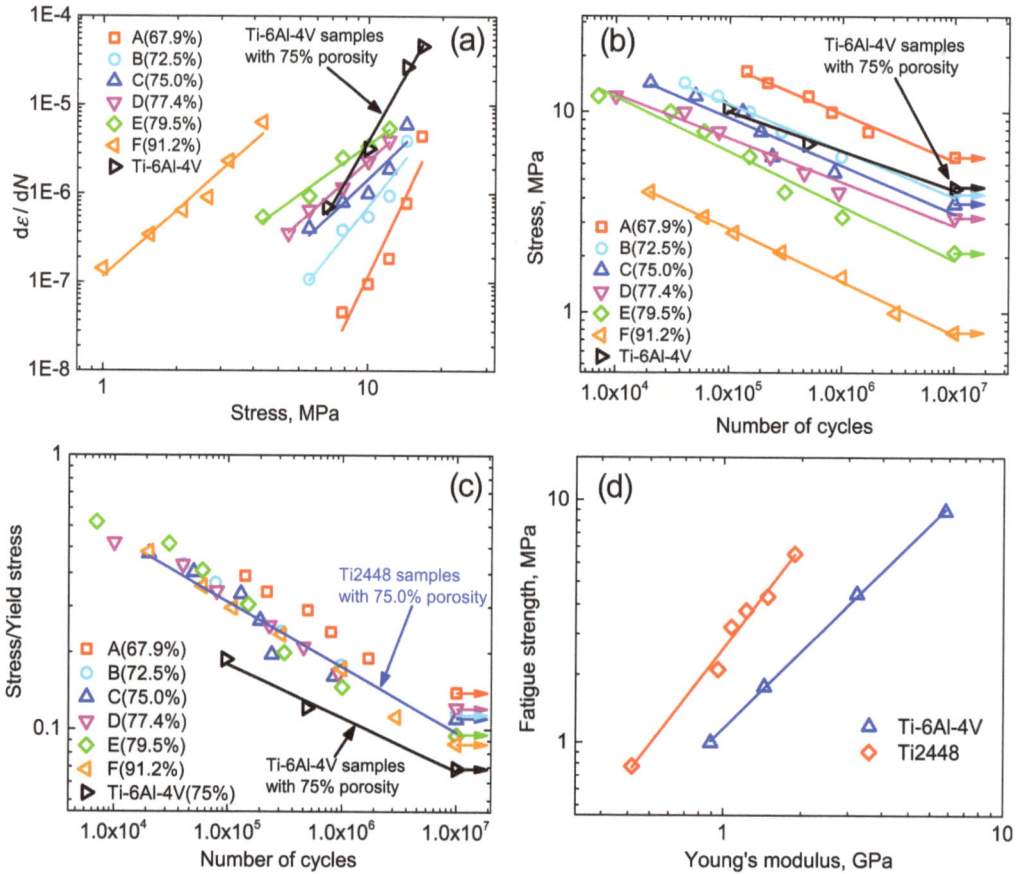

Fig. (7). (a) The cyclic ratcheting rate of the porous Ti2448 and Ti-6Al-4V specimens with different porosity, the S-N curves **(b)** and the normalized S-N curves **(c)** of the porous Ti2448 and Ti-6Al-4V specimens with different porosity, and **(d)** the relationship of Young's modulus and the fatigue strength for the porous Ti2448 and Ti-6Al-4V specimens.

Fatigue Crack Morphology and Deflection

The fatigue crack surfaces of the porous samples were observed using SEM and micro-CT technologies. Fig. **(8a)** reveals that most cracks occur on the node of the porous structures, and the micro-CT image (Fig. **8b**) displays that crack initiation generates at the top of the node, which is subjected tensile stress [36]. The results of EBSD analysis on the fatigue crack propagation track for the porous sample (67.9%) are plotted in Fig. **(9)**. The β grains including the columnar grains and equiaxed grains are clearly surrounded by the crack tracks. Although the crack propagation presents transgranular fracture, the fatigue cracks deflect at the high angle grain boundaries of the columnar grains instead of passing through grain boundaries. Such crack deflections do not occur in equiaxed grains.

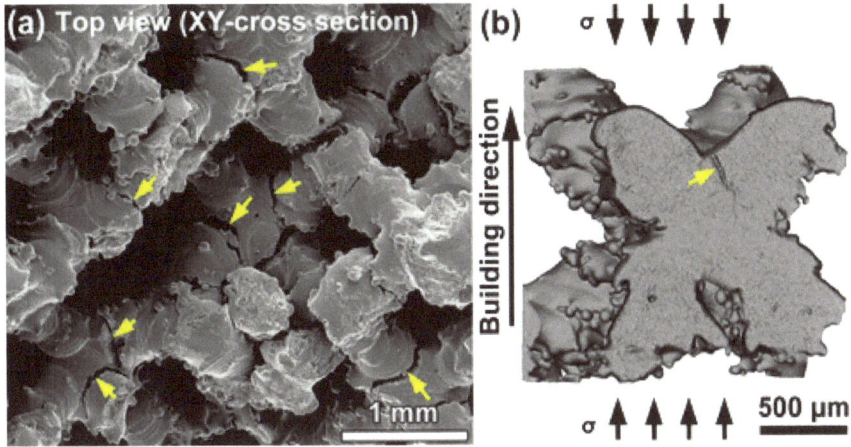

Fig. (8). (**a**) The top view of the rhombic dodecahedron structure after fatigue test, (**b**) the position of the fatigue crack initiation in micro-CT image.

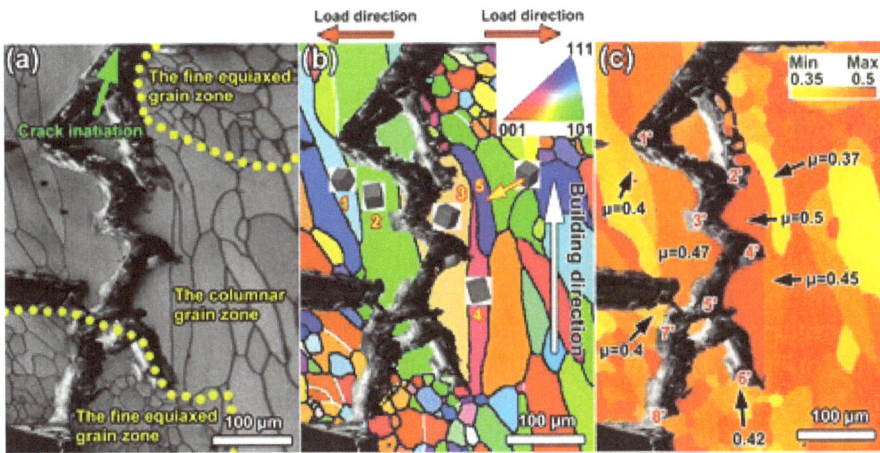

Fig. (9). The EBSD image of the fatigue crack propagation track for the porous Ti2448 specimen with the porosity of 67.9%, (**a**) the morphology of the single β grains, (**b**) the EBSD orientation microscopy map of the strut along the vertical direction, and (**c**) the EBSD Schmid factor mapping image along (110) <111>β direction.

Recent studies showed that the microstructural characteristics including constituent phases, grain size and shape, grain boundaries, and the misorientation angle between adjacent grains play significant roles in influencing fatigue properties [46 - 48], which are related to the fatigue crack track and the propagation rate. The β grains are composed of fine equiaxed grains and coarse columnar grains (Fig. **9**). As the manufacturing process is a unique powder rapid melting and solidification process in additive manufacturing, it is therefore unavoidably some powder, which distributes at the edge of the melt pool and receives relatively less input energy, would subsequently adhere to the strut surface during building process [49] (most of these adhered powder could be removed in powder removing system (PRS)). These adhered powder acts as heterogeneous nucleation sites for β grains thereby affecting their surrounding microstructure to form the fine equiaxed grain zone. The formation mechanism of the fine equiaxed grain zone is similar to the chill zone in the cast sample, caused by the heterogeneous nucleation occurring on strut surface [50], whereas the formation of coarse columnar grains is mainly determined by the temperature gradient during EBM process [44, 51].

From a microscopic point of view, the fatigue crack in this work indicates significant deflection in both the fine equiaxed grain zone and the coarse columnar grain zone, and interestingly, the positions of the deflection locate at the grain boundaries [52]. The misorientation between the grains is more than 15° (Fig. **9b**). Clearly the fatigue cracks do not easily propagate through the grain boundaries at some places, such as the points 1', 2', 3', 4', 5' and 7' in Fig. (**9c**). It is reported that short cracks could form along a slip plane within a grain and that the misorientation between the adjacent grains could influence the crack path significantly [48]. Similarly, the large misorientation between the adjacent grains, such as grains 1 and 2 in Fig. (**9b**), would block the short cracks to propagate through grain boundaries.

In order to fully understand the crack deflection effect, the Schmid factor mapping image along (110) $<111>_\beta$ direction is plotted in Fig. (**9c**) to analyze this effect. In general, the Schmid factor (μ) can be used to examine the plastic deformation [47]. A higher value of the Schmid factor indicates that the grain is more favorably oriented for slip. Clearly, most β grains in the sample present a large Schmid factor in the range of 0.37-0.5. It is well known that the slip bands can evolve from the dislocations and then transfer to the microcracks [48]. As the size of strut thickness is quite small, the fatigue properties are mainly affected by the short fatigue crack propagation mechanism in the single β phase titanium structure. The estimation of plastic zone size (r_p) ahead of the fatigue crack tip can be determined by Eq. (3) [53],

$$r_p = (k_{max}/\sigma_y)^2/2\pi \qquad\qquad (3)$$

where K_{max} is the maximum applied stress intensity factor and σ_y is the yield strength. Based on our previous study [54], the size of plastic zone (r_p) is ~187 μm, which exceeds the columnar grain size (with a width of less than 100 μm) thereby activating more than one slip system within the neighboring grains [55]. The orientation of the grains near the crack presents a large difference (Fig. **9**), which results in the crack propagation direction deflecting along a favored slip system, leading to a crack deflection effect. In particular, the obvious crack deflection can be found in long columnar grain. Here, it can be seen that the dislocations cannot transfer into the adjacent grain with the low Schmid factor. Then the slip bands tend to return back at the grain boundaries, thereby deflecting the main crack at the grain boundaries where the neighboring grains have the low Schmid factor. It is found that all the deflection positions in this chapter conform to this mechanism; they are the points 1', 2', 3', 4', 5' 7'and 8'.

COMPARISON OF FATIGUE BETWEEN TI2448 AND TI6-AL-4V POROUS STRUCTURES

Generally, the normalized fatigue strengths of the porous structures are lower than those for their bulk counterparts. This is caused by the rough surface and the defects of the strut, which could improve the crack initiation generation thereby reducing the fatigue life. It was reported that the fatigue properties of Ti-6Al-4V could be improved successfully by changing the unit shape for porous structures. The improvement in fatigue properties is mainly realized through balancing the bending and buckling stress to reduce the stress concentration, thereby optimizing the ratcheting effect, delaying the crack initiation and reducing the crack propagation rate. It is usually believed that the fatigue properties could be improved by increasing the material toughness for bulk samples [56]. Similarly, this should be suitable for a given unit shape structure. Thus, theoretically, the single β phase Ti2448 alloy with larger ductility and lower Young's modulus would sustain a better fatigue performance than Ti-6Al-4V.

The interaction of the ratcheting effect and the fatigue damage mechanism should strongly influence the fatigue properties for the titanium porous samples. The cyclic ratcheting effect is the accumulation of cyclic plastic deformation during the fatigue process, which is determined by the applied stress level and the ductility of the material [37]. Better ductility would lead to a lower cyclic ratcheting rate. This chapter shows that the single β Ti2448 structures for all porosity levels present super-elastic property and excellent ductility (more than 30%). By contrast, the ductility of Ti-6Al-4V porous structure with 75% porosity is less than 10%. Consequently, the cyclic ratcheting rates of Ti2448 samples ($d\varepsilon/dN$) are much lower than that of Ti-6Al-4V samples (Fig. **7a**).

Furthermore, the fatigue damage effect, which refers to the fatigue crack propagation rate, plays a dominant role in fatigue properties. Indeed, two aspects could affect the fatigue damage effect, *i.e.* the crack initiation and the crack propagation process. Crack initiation normally relies on the surface roughness and applied stress level. As the EBM-produced samples with similar surface roughness, the Ti-6Al-4V and Ti2448 porous samples are mainly affected by the stress level. That is, the high stress level results in much earlier crack initiation than the low stress level. The crack propagation rate depends largely on the material properties and the stress level. The behavior of the crack propagation in porous structures differs slightly from to that of solid bulk samples. The material properties, including the strength, toughness, Young's modulus as well as the grain size, can affect the crack propagation rate. During the fatigue process, dislocation pile-ups would form at the plastic zone ahead of the crack tip. Then those dislocation pile-ups evolve into the slip bands as the fatigue cycles increase. It can be seen that the grain boundaries can stop the spread of dislocation pile-ups. So for the Ti2448 samples with the large single β grain (more than 100 μm), the plastic zone ahead of the crack tip is quite large. By contrast, the dislocations in the Ti-6Al-4V samples concentrate at the interfaces of the α' phase with the size less than 100 nm, so the plastic zone ahead of the crack tip is much smaller than for Ti2448. Thus, the stress concentration ahead of the fatigue crack tip is high and the crack propagation rate in the Ti-6Al-4V samples is much faster than in the Ti2448 samples. Furthermore, the deflection of the crack propagation would increase the toughness therefore improve the fatigue properties [57, 58]. In summary, the Ti2448 porous samples with higher toughness and larger plastic zone ahead of the crack tip lead to a lower fatigue crack propagation rate and a higher fatigue life than Ti-6Al-4V porous samples.

CONCLUDING REMARKS

In this chapter, the manufacturing process, microsturcutre and mechanical properties of the single β phase Ti-24Nb-4Zr-8Sn porous samples with different porosity levels were studied systematically. The variation of super-elastic property, Young's modulus and fatigue properties, and the fatigue crack deflection behavior were detailed investigated. The super-elastic property can be improved by increasing porosity of porous samples as a result of increasing the ratio of tensile/compressive of the porous structure. The position of fatigue crack initiation on the strain curves is defined depend on the variation of the fatigue cyclic loops. The unique manufacturing process of the EBM results in formation of different sizes of grains. Therefore, the apparent fatigue crack deflection occurs in the columnar grain zone due to large misorientation between the adjacent grains. Compared with Ti-6Al-4V samples, the Ti-24Nb-4Zr-8Sn porous samples exhibit a higher normalized fatigue strength owing to the super-elastic property and the larger plastic zone ahead of the fatigue crack tip. For the same fatigue strength, the Young's modulus of Ti-24Nb-4Zr-8Sn porous samples is only half of Ti-6Al-4V porous samples.

CONFLICT OF INTEREST

The author (editor) declares no conflict of interest, financial or otherwise.

ACKNOWLEDGEMENTS

This research was supported under the Australian Research Council's Discovery Projects (DP110101653). The authors are grateful to S.J. Li, T.B Sercombe, and Y.L. Hao for collaborations.

REFERENCES

[1] Kurtz SM, Ong KL, Schmier J, *et al.* Future clinical and economic impact of revision total hip and knee arthroplasty. J Bone Joint Surg Am 2007; 89 (Suppl. 3): 144-51.
[PMID: 17908880]

[2] Liu Y, Zhao X, Zhang L-C, Habibi D, Xie Z. Architectural design of diamond-like carbon coatings for long-lasting joint replacements. Mater Sci Eng C 2013; 33(5): 2788-94.
[http://dx.doi.org/10.1016/j.msec.2013.02.047] [PMID: 23623097]

[3] Peltola SM, Melchels FP, Grijpma DW, Kellomäki M. A review of rapid prototyping techniques for tissue engineering purposes. Ann Med 2008; 40(4): 268-80.
[http://dx.doi.org/10.1080/07853890701881788] [PMID: 18428020]

[4] Zhang L-C, Attar H. Selective Laser Melting of Titanium Alloys and Titanium Matrix Composites for Biomedical Applications: A Review. Adv Eng Mater 2016; 18: 463-75.
[http://dx.doi.org/10.1002/adem.201500419]

[5] Koike M, Greer P, Owen K, *et al.* Evaluation of Titanium Alloys Fabricated Using Rapid Prototyping Technologies-Electron Beam Melting and Laser Beam Melting. Materials (Basel) 2011; 4(10): 1776-92.
[http://dx.doi.org/10.3390/ma4101776] [PMID: 28824107]

[6] Murr LE, Gaytan S, Ceylan A, *et al.* Characterization of titanium aluminide alloy components fabricated by additive manufacturing using electron beam melting. Acta Mater 2010; 58: 1887-94.
[http://dx.doi.org/10.1016/j.actamat.2009.11.032]

[7] Sercombe TB, Xu X, Challis VJ, *et al.* Failure modes in high strength and stiffness to weight scaffolds produced by Selective Laser Melting. Mater Des 2015; 67: 501-8.
[http://dx.doi.org/10.1016/j.matdes.2014.10.063]

[8] Liu YJ, Li XP, Zhang L-C, Sercombe TB. Processing and properties of topologically optimised biomedical Ti-24Nb-4Zr-8Sn scaffolds manufactured by selective laser melting. Mater Sci Eng A 2015; 642: 268-78.
[http://dx.doi.org/10.1016/j.msea.2015.06.088]

[9] Tsirkas SA, Papanikos P, Kermanidis T. Numerical simulation of the laser welding process in butt-joint specimens. J Mater Proc Tech 2003; 134: 59-69.
[http://dx.doi.org/10.1016/S0924-0136(02)00921-4]

[10] Liu Y, Wang W, Zhang L-C. Additive manufacturing techniques and their biomedical applications. Family Med Comm Health 2017.

[11] Geetha M, Singh A, Asokamani R, Gogia A. Ti based biomaterials, the ultimate choice for orthopaedic implants-a review. Prog Mater Sci 2009; 54: 397-425.
[http://dx.doi.org/10.1016/j.pmatsci.2008.06.004]

[12] Long M, Rack HJ. Titanium alloys in total joint replacement--a materials science perspective. Biomaterials 1998; 19(18): 1621-39.

[http://dx.doi.org/10.1016/S0142-9612(97)00146-4] [PMID: 9839998]

[13] Bai Y, Gai X, Li S, *et al.* Improved corrosion behaviour of electron beam melted Ti-6Al-4 V alloy in phosphate buffered saline. Corros Sci 2017; 123: 289-96.
[http://dx.doi.org/10.1016/j.corsci.2017.05.003]

[14] Dai N, Zhang L-C, Zhang J, *et al.* Distinction in Corrosion Resistance of Selective Laser Melted Ti-6Al-4V Alloy on Different Planes. Corros Sci 2016; 111: 703-10.
[http://dx.doi.org/10.1016/j.corsci.2016.06.009]

[15] Challis VJ, Xu X, Zhang L-C, Roberts AP, Grotowski JF, Sercombe TB. High specific strength and stiffness structures produced using selective laser melting. Mater Des 2014; 63: 783-8.
[http://dx.doi.org/10.1016/j.matdes.2014.05.064]

[16] Haghighi SE, Lu H, Jian G, Cao G, Habibi D, Zhang L-C. Effect of α″martensite on the microstructure and mechanical properties of beta-type Ti-Fe-Ta alloys. Mater Des 2015; 76: 47-54.
[http://dx.doi.org/10.1016/j.matdes.2015.03.028]

[17] Zhang L-C, Klemm D, Eckert J, Hao YL, Sercombe TB. Manufacture by selective laser melting and mechanical behavior of a biomedical Ti-24Nb-4Zr-8Sn alloy. Scr Mater 2011; 65: 21-4.
[http://dx.doi.org/10.1016/j.scriptamat.2011.03.024]

[18] Dai N, Zhang L-C, Zhang J, Chen Q, Wu M. Corrosion behavior of selective laser melted Ti-6Al-4V alloy in NaCl solution. Corros Sci 2016; 102: 484-9.
[http://dx.doi.org/10.1016/j.corsci.2015.10.041]

[19] Hao YL, Yang R, Niinomi M, *et al.* Aging response of the young's modulus and mechanical properties of Ti-29Nb-13Ta-4.6Zr for biomedical applications. Metall Mater Trans, A Phys Metall Mater Sci 2003; 34: 1007-12.
[http://dx.doi.org/10.1007/s11661-003-0230-x]

[20] Liu Y, Wang H, Li S, *et al.* Compressive and fatigue behavior of beta-type titanium porous structures fabricated by electron beam melting. Acta Mater 2017; 126: 58-66.
[http://dx.doi.org/10.1016/j.actamat.2016.12.052]

[21] Hao YL, Li SJ, Sun SY, Zheng CY, Hu QM, Yang R. Super-elastic titanium alloy with unstable plastic deformation. Appl Phys Lett 2005; 87: 091906.
[http://dx.doi.org/10.1063/1.2037192]

[22] Liu YJ, Li SJ, Hou WT, *et al.* Electron beam melted beta-type Ti-24Nb-4Zr-8Sn porous structures with high strength-to-modulus ratio. J Mater Sci Technol 2016; 32: 505-8.
[http://dx.doi.org/10.1016/j.jmst.2016.03.020]

[23] Hrabe NW, Heinl P, Flinn B, Körner C, Bordia RK. Compression-compression fatigue of selective electron beam melted cellular titanium (Ti-6Al-4V). J Biomed Mater Res B Appl Biomater 2011; 99(2): 313-20.
[http://dx.doi.org/10.1002/jbm.b.31901] [PMID: 21948776]

[24] Hernandez J, Li S, Martinez E, *et al.* Microstructures and Hardness Properties for β-Phase Ti-24N--4Zr-7.9 Sn Alloy Fabricated by Electron Beam Melting. J Mater Sci Technol 2013; 29: 1011-7.
[http://dx.doi.org/10.1016/j.jmst.2013.08.023]

[25] Simchi A. Direct laser sintering of metal powders: Mechanism, kinetics and microstructural features. Mater Sci Eng A 2006; 428: 148-58.
[http://dx.doi.org/10.1016/j.msea.2006.04.117]

[26] Semak V, Matsunawa A. The role of recoil pressure in energy balance during laser materials processing. J Phys D Appl Phys 1998; 30: 2541-52.
[http://dx.doi.org/10.1088/0022-3727/30/18/008]

[27] Gong H, Rafi K, Gu H, Ram GD, Starr T, Stucker B. Influence of defects on mechanical properties of Ti-6Al-4V components produced by selective laser melting and electron beam melting. Mater Des 2015; 86: 545-54.

[http://dx.doi.org/10.1016/j.matdes.2015.07.147]

[28]   Lemasson P, Carin M, Parpillon JC, Berthet R. 2D-heat transfer modelling within limited regions using moving sources: application to electron beam welding. Int J Heat Mass Transfer 2003; 46: 4553-9.
[http://dx.doi.org/10.1016/S0017-9310(03)00288-6]

[29]   Dowden JM. The mathematics of thermal modeling: an introduction to the theory of laser material processing. Boca Raton: CRC Press 2001.
[http://dx.doi.org/10.1201/9781420035629]

[30]   Geiger M, Leitz KH, Koch H, Otto A. A 3D transient model of keyhole and melt pool dynamics in laser beam welding applied to the joining of zinc coated sheets. Prod Eng 2009; 3: 127-36.
[http://dx.doi.org/10.1007/s11740-008-0148-7]

[31]   Tian Y, Wang C, Zhu D, Zhou Y. Finite element modeling of electron beam welding of a large complex Al alloy structure by parallel computations. J Phys D Appl Phys 2008; 199: 41-8.

[32]   Courtois M, Carin M, Masson PL, Gaied S, Balabane M. A new approach to compute multi-reflections of laser beam in a keyhole for heat transfer and fluid flow modelling in laser welding. J Phys D Appl Phys 2013; 46: 505305.
[http://dx.doi.org/10.1088/0022-3727/46/50/505305]

[33]   Tang Q, Pang S, Chen B, Suo H, Zhou J. A three dimensional transient model for heat transfer and fluid flow of weld pool during electron beam freeform fabrication of Ti-6Al-4V alloy. Int J Heat Mass Transfer 2014; 78: 203-15.
[http://dx.doi.org/10.1016/j.ijheatmasstransfer.2014.06.048]

[34]   Yavari SA, Wauthlé R, van der Stok J, *et al.* Fatigue behavior of porous biomaterials manufactured using selective laser melting. Mater Sci Eng C 2013; 33(8): 4849-58.
[http://dx.doi.org/10.1016/j.msec.2013.08.006] [PMID: 24094196]

[35]   Li SJ, Xu QS, Wang Z, *et al.* Influence of cell shape on mechanical properties of Ti-6Al-4V meshes fabricated by electron beam melting method. Acta Biomater 2014; 10(10): 4537-47.
[http://dx.doi.org/10.1016/j.actbio.2014.06.010] [PMID: 24969664]

[36]   Chen LF, Zhao PF, Xie HQ, Yu W. Thermal properties of epoxy resin based thermal interfacial materials by filling Ag nanoparticle-decorated graphene nanosheets. Comp Sci Tech 2016; 125: 17-21.
[http://dx.doi.org/10.1016/j.compscitech.2016.01.011]

[37]   Li S, Murr L, Cheng X, *et al.* Compression fatigue behavior of Ti-6Al-4V mesh arrays fabricated by electron beam melting. Acta Mater 2012; 60: 793-802.
[http://dx.doi.org/10.1016/j.actamat.2011.10.051]

[38]   Ehtemam-Haghighi S, Liu Y, Cao G, Zhang L-C. Influence of Nb on the β→α″ martensitic phase transformation and properties of the newly designed Ti-Fe-Nb alloys. Mater Sci Eng C 2016; 60: 503-10.
[http://dx.doi.org/10.1016/j.msec.2015.11.072] [PMID: 26706557]

[39]   Ehtemam-Haghighi S, Prashanth KG, Attar H, Chaubey AK, Cao GH, Zhang L-C. Evaluation of mechanical and wear properties of Ti-x-Nb 7Fe alloys designed for biomedical applications. Mater Des 2016; 11: 592-9.
[http://dx.doi.org/10.1016/j.matdes.2016.09.029]

[40]   Ehtemam-Haghighi S, Liu Y, Cao G, Zhang L-C. Phase transition, microstructural evolution and mechanical properties of Ti-Nb-Fe alloys induced by Fe addition. Mater Des 2016; 97: 279-86.
[http://dx.doi.org/10.1016/j.matdes.2016.02.094]

[41]   Zhang SQ, Li SJ, Jia M, *et al.* Low-cycle fatigue properties of a titanium alloy exhibiting nonlinear elastic deformation behavior. Acta Mater 2011; 59: 4690-9.
[http://dx.doi.org/10.1016/j.actamat.2011.04.015]

[42]   Ashby MF, Evans T, Fleck NA, Hutchinson J, Wadley H, Gibson L. Metal foams: a design guide.

Elsevier 2000.

[43]   Özbilen S, Liebert D, Beck T, Bram M. Fatigue behavior of highly porous titanium produced by powder metallurgy with temporary space holders. Mater Sci Eng C 2016; 60: 446-57.
[http://dx.doi.org/10.1016/j.msec.2015.11.050] [PMID: 26706551]

[44]   Liu YJ, Li SJ, Wang HL, *et al.* Microstructure, defects and mechanical behavior of beta-type titanium porous structures manufactured by electron beam melting and selective laser melting. Acta Mater 2016; 113: 56-67.
[http://dx.doi.org/10.1016/j.actamat.2016.04.029]

[45]   Lefebvre L-P, Baril E, Bureau MN. Effect of the oxygen content in solution on the static and cyclic deformation of titanium foams. J Mater Sci Mater Med 2009; 20(11): 2223-33.
[http://dx.doi.org/10.1007/s10856-009-3798-x] [PMID: 19554427]

[46]   Kamp N, Gao N, Starink M, Sinclair I. Influence of grain structure and slip planarity on fatigue crack growth in low alloying artificially aged 2xxx aluminium alloys. Int J Fatigue 2007; 29: 869-78.
[http://dx.doi.org/10.1016/j.ijfatigue.2006.08.005]

[47]   Wei L, Pan Q, Huang H, Feng L, Wang Y. Influence of grain structure and crystallographic orientation on fatigue crack propagation behavior of 7050 alloy thick plate. Int J Fatigue 2014; 66: 55-64.
[http://dx.doi.org/10.1016/j.ijfatigue.2014.03.009]

[48]   Zhai T, Wilkinson A, Martin J. A crystallographic mechanism for fatigue crack propagation through grain boundaries. Acta Mater 2000; 48: 4917-27.
[http://dx.doi.org/10.1016/S1359-6454(00)00214-7]

[49]   Khairallah SA, Anderson AT, Rubenchik A, King WE. Laser powder-bed fusion additive manufacturing: Physics of complex melt flow and formation mechanisms of pores, spatter, and denudation zones. Acta Mater 2016; 108: 36-45.
[http://dx.doi.org/10.1016/j.actamat.2016.02.014]

[50]   Phulé PP. Essentials of materials science and engineering. Tsinghua University Press 2005.

[51]   Tan X, Kok Y, Tan YJ, *et al.* Graded microstructure and mechanical properties of additive manufactured Ti-6Al-4V *via* electron beam melting. Acta Mater 2015; 97: 1-16.
[http://dx.doi.org/10.1016/j.actamat.2015.06.036]

[52]   Chen YQ, Pan SP, Zhou MZ, Yi DQ, Xu DZ, Xu YF. Effects of inclusions, grain boundaries and grain orientations on the fatigue crack initiation and propagation behavior of 2524-T3 Al alloy. Mater Sci Eng A 2013; 580: 150-8.
[http://dx.doi.org/10.1016/j.msea.2013.05.053]

[53]   Suresh S, Ritchie RO. Propagation of short fatigue cracks. Int Mater Rev 1983; 29: 445-75.
[http://dx.doi.org/10.1179/imr.1984.29.1.445]

[54]   Zhang SQ. Investigation of fatigue behaviour of Ti2448 alloy. Shenyang: Institute of Metal Research,Chinese Academy of Sciences 2010.

[55]   Suresh S, Ritchie R. A geometric model for fatigue crack closure induced by fracture surface roughness. Metall Trans, A, Phys Metall Mater Sci 1982; 13: 1627-31.
[http://dx.doi.org/10.1007/BF02644803]

[56]   Ritchie RO. The conflicts between strength and toughness. Nat Mater 2011; 10(11): 817-22.
[http://dx.doi.org/10.1038/nmat3115] [PMID: 22020005]

[57]   Evans AG. Perspective on the development of high-toughness ceramics. J Am Ceram Soc 1990; 73: 187-206.
[http://dx.doi.org/10.1111/j.1151-2916.1990.tb06493.x]

[58]   Ritchie R. Mechanisms of fatigue crack propagation in metals, ceramics and composites: role of crack tip shielding. Mater Sci Eng A 1988; 103: 15-28.
[http://dx.doi.org/10.1016/0025-5416(88)90547-2]

CHAPTER 6

Preparation, Microstructure and Mechanical Properties of NiTi-Nb Porous Titanium Alloy

Liqiang Wang[1,*], Wei Huang[1], Wei Zhang[1] and Lai-Chang Zhang[2]

[1] *State Key Laboratory of Metal Matrix Composites, Shanghai Jiao Tong University, No. 800 Dongchuan Road, Shanghai200240, PR China*

[2] *School of Engineering, Edith Cowan University, 270 Joondalup Drive, Joondalup, Perth, WA, 6027, Australia*

Abstract: The excellent mechanical properties of NiTi alloys, such as unique shape memory effect, superelasticity, as well as good biocompatibility, excellent corrosion resistance, low elastic modulus and good ductility, make it an ideal choice for the biomedical, aerospace and intelligent materials. However, NiTi alloy with itself and dissimilar material connection problems NiTi alloy can only have a simple geometry, thus limiting the NiTi alloy more widely used. For this reason, resistance welding and laser welding are proposed by scholars, but these methods are not widely used due to the fact that welding makes the joint parts easy to produce brittle second phase and large area heat-affected zone, which also needs a large amount of solder. In this chapter, NiTi / NiTi-Nb alloy was fabricated by heating niobium and NiTi wires in a high-temperature argon atmosphere furnace to 1185 °C for 6min. The results show that there are 4 diffusion layers, including Nb foil, NiTi-Nb eutectic region, pre-eutectic NiTi region and NiTi matrix at the junction of NiTi wire and Nb foil. In addition, martensite phase was found inside the NiTi matrix. In the NiTi-Nb eutectic diffusion layer, niobium is inhomogeneously distributed, forming a rod-like Nb-rich phase, facetted Ti-rich phase and stripe-like or equiaxed NiTi-Nb eutectic structure. The experimental study on the microstructure and mechanical properties of interface interface between NiTi and Nb will help to understand the advantages of using Nb as the NiTi alloy connection material, provide theoretical support for preparation of NiTi lattice materials, and promote NiTi alloy more extensive application.

Keywords: Diffusion layers, Mechanical property, Microstructure, NiTi/NiTi-Nb alloys.

** **Corresponding author Liqiang Wang:** State Key Laboratory of Metal Matrix Composites, School of Materials Science and Engineering, Shanghai Jiao Tong University, No. 800 Dongchuan Road, Shanghai 200240, P.R. China; Tel: 8602134202641; Fax: 8602134202749; E-mail: wang_liqiang@sjtu.edu.cn*

INTRODUCTION

Shape Memory Alloy

Shape Memory Alloys (SMA) can recover their original shape from a significant deformation when an appropriate stimulus is applied [1]. In the 1930s, the shape memory effect (SME) is firstly discovered by Ölander in the Cd-Au system. The study found the alloy with a plastic deformation of up to 47.5% under low temperature can recover its original size and shape after heat treatment when Cd content is less than 50% [2]. Ten years later, Kurdjumov and Khandros firstly proposed that thermoelastic martensitic transformation is the direct cause of SME [3]. During the following decades, SME has also been found in other binary alloy systems and ternary alloy systems, such as Cu-Zn, Tl-In and Cu-Al-Ni [4 - 6].

Until 1959, the NiTi alloy is firstly discovered by William J. Buehler in the US Navy ordnance laboratory [7]. Three years later, William J. Buehler and Frederick E. Wang found that NiTi alloys also had significant recoverable deformability [8]. At the same time, due to its excellent biocompatibility and fine comprehensive mechanical properties, the commercial value of NiTi alloys is recognized in many fields, especially in the biomedical field, such as bone graft stents and surgical guide wires. It's also widely applied in the fields of robot, automobile, aerospace and intelligent materials [9 - 11].

Today, a variety of shape memory alloys have been developed, such as solid, thin films and even foam materials. Nickel-based, copper-based and iron-based shape memory alloys have attracted considerable concern in the business community. According to the key performance indicators in engineering application, experts and scholars compared these three typical shape memory alloy systematically (shown in Table 1). The results show that NiTi-based SMAs should be the first choice for engineering materials due to the excellent mechanical performances and biocompatibility [12]. Compared with NiTi-based alloys, Cu-based shape memory alloys have cost advantages and good processability. In addition, some Cu-based alloys in martensitic state have deformation recovery performance after aging treatment similar to rubber [13]. The Fe-based alloys have relatively weak shape memory effect. They are applied to a handful of fastening device of micro-drive equipment merely because of the low cost [14].

NiTi and NiTiNb Alloy

Phase Transition of NiTi Alloy

As shown in Fig. (1), the ordered intermediate compound of NiTi forms when the composition is near 50 at.pct Ni in Ni-Ti system. In fact, the system is in non-equilibrium state due to the fast heating or cooling rates, therefore equiatomic Ni-Ti alloy could form with composition of 48-52 at% Ti and a few other compounds usually occur at the room temperature. At room temperature, there are two other stable binary Ni-Ti compounds of Ni-rich Ni_3Ti and Ti-rich Ti_2Ni, respectively. These compounds are usually produced during processing and it is difficult to remove them completely due to their good stability, therefore gaining high-purified equiatomic NiTi alloy. The Ti_2Ni phase precipitation problem is especially serious because the brittle Ti_2Ni phase forms near the grain boundary in the process of powder metallurgy, which deteriorates plasticity and toughness of NiTi alloy. For Ni-rich near-equiatomic Ni-Ti alloy ($50< x < 51.8$ at %), Ni_4Ti_3 precipitates in the aging process, which can improve the superelastic properties of NiTi alloy. Therefore, in order to optimize the superelastic performance, the aging treatment is needed after powder metallurgy processing [15].

Table 1. Property indicators of typical shape memory alloys.

Performance index	NiTi	CuZnAl	CuAlNi
Thermal conducticvity(20°C)(W/mK)	8.6-18	84-120	30-75
density(Kg/m³)	6400-6500	7540-8000	7100-7200
Specific resistance(10^6Ům)	0.5-1.1	0.07-0.12	0.1-0.14
Thermal expansivity(10-6/K)	6.6-11	17	17
Maximum recovery stress(Mpa)	500-900	400-700	300-600
Normal working stress(Mpa)	100-130	40	70
Fatigue strength(N=10^6)(Mpa)	350	270	350
Young modulus(Gpa)	28-83	70-100	80-100
Shape memory transition temperature(°C)	-200-200	-200-150	-200-200
Heat stagnation(°C)	2-50	5-20	20-40
Damping property(SDC%)	15-20	30-85	10-20
Grain size(μm)	1-100	50-150	25-100
formability(rolling)	difficult	easy	difficult
Smelting, casting and composition control	difficult	general	general

The martensitic phase transition can occur between two different crystal structures of the binary NiTi alloy *via* heating (cooling) or plastic deformation. In the relatively high temperature condition, NiTi alloy is austenite phase and its crystal structure is body-centered cubic B2 structure. However, in the relatively low temperature condition, NiTi alloy is martensitic phase and its crystal structure is monoclinic B19' structure. The crystal structure image of austenite phase B2 and martensite phase B19' are shown in Fig. (**2**). The R phase can be formed in a variety of ways, such as thermal cycling, cold working and aging treatment.

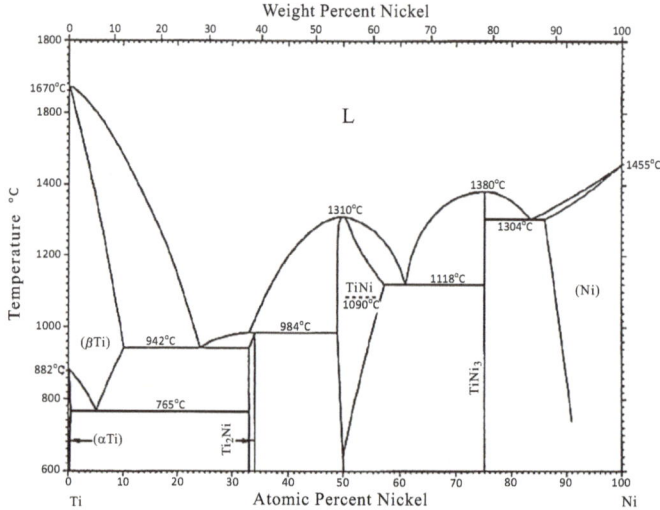

Fig. (1). Ni-Ti binary phase diagram [16].

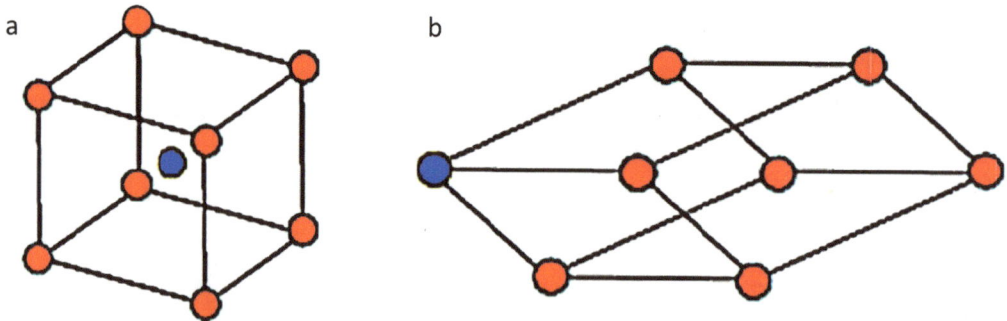

Fig. (2). NiTi structure: (**a**) austenite B2 structure; (**b**) martensite B19' structure.

According to the phase transition energy theory, the martensitic transformation occurs when the Gibbs free energy of austenite is higher than that of martensite. However, the martensitic transformation does not occur at a fixed temperature without a given stress and requires a certain condensate depression as a driving force to promote the transition from austenite phase to martensite phase. Usually A_S, A_F express the austenite transformation initiating temperature and the austenite transformation finished temperature, respectively. Similarly, M_S, M_F express the martensite transformation initiating temperature and the martensite transformation finished temperature, respectively. During the cooling process from austenite phase, the martensitic transition temperature range is from M_S to M_F. During the martensite phase heating process, the austenitic transition temperature range is from A_S to A_F. The phase transition temperatures of A_S, A_F, M_S and M_F can be measured by differential scanning calorimetry (DSC). In the temperature-heat discharge curve, there will be heat peaks when the martensitic transformation or its reverse transformation occurs, as shown in Fig. (**3**).

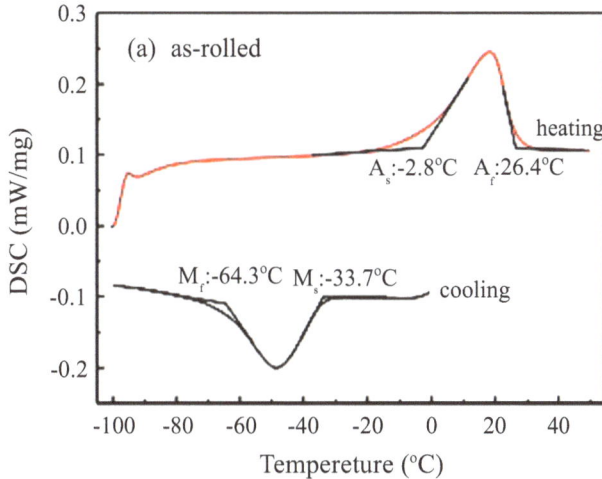

Fig. (3). Schematic diagram of DSC curve.

The martensitic phase transformation and its reversal transformation temperature mainly depend on the chemical composition and processing technology. In general, for the Ti-rich phase, the austenite transition temperature is relatively high. For the Ni-rich phase, the martensitic transition temperature is relatively low. Therefore, for NiTi alloy, the working temperature should be considered seriously because the temperature will affect the microstructure of NiTi alloy significantly, and even its SME and superelastic property.

The transition between the austenite phase and the martensite phase of the NiTi alloy is non-diffusion and solid phase transformation and usually goes on in the shear mode of the crystal face, the structure transition schematic diagram of martensite transformation shown in Fig. (**4**). The B2 phase plane (110) corresponds to the martensitic phase plane (001). It is generally believed that martensite transformation occurs *via* the B2 phase lattice plane (110) compresses along the direction of the B2 phase crystal orientation [001] and stretches in the direction of the B2 phase crystal orientation [001]. Meanwhile, the B2 phase lattice plane (110) shear in direction of the B2 phase crystal orientation [110] with another lattice plane. Thus, martensite phases with twin boundaries form. Both temperature and stress can induce martensite transformation and its internal transformation. The relationship between phase transformation, temperature and stress can be expressed by the Krause-Krappel equation [17].

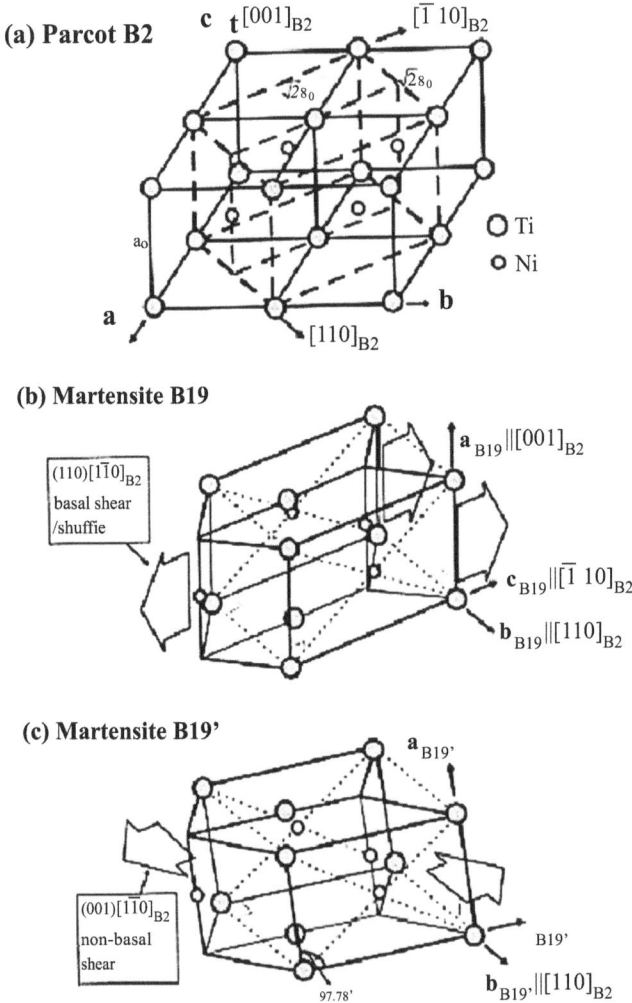

Fig. (4). Structure relationships between B2, B19 and B19'. (**a**) B2 parent phase structure; (**b**) martensite B19 structure; (**c**) martensite B19' structure.

$$\frac{d\sigma}{dT} = -\frac{\rho \Delta S}{\varepsilon} = -\frac{\rho \Delta H}{T_0 \varepsilon} \tag{1}$$

Where ρ is the material density(kg/cm^3), ΔS is the entropy change(J/(mol·K)), ΔH is the enthalpy change(J/(mol), T_0 is the equilibrium phase transition temperature(K), ε is strain in the direction of stress and σ is stress(N/m^2).

NiTi Alloy Properties

(1) Shape Memory Effect

For Ti-rich quasi-equiatomic Ti-Ni alloy, martensitic phase appears at room temperature. And its microstructure contains many self-accommodated martensite variants. Since shape memory alloy phase transition performs from the high symmetry of austenite phase to the low symmetry of martensite phase, it can produce 24 identical crystal structure variants. These variants are all twins and appear in pairs in order to maintain no shape change during phase transformation. Therefore, self-accommodated martensites form and the detailed explanation shows typical stress-strain-temperature curve of shape memory alloys in Fig. (**5**).

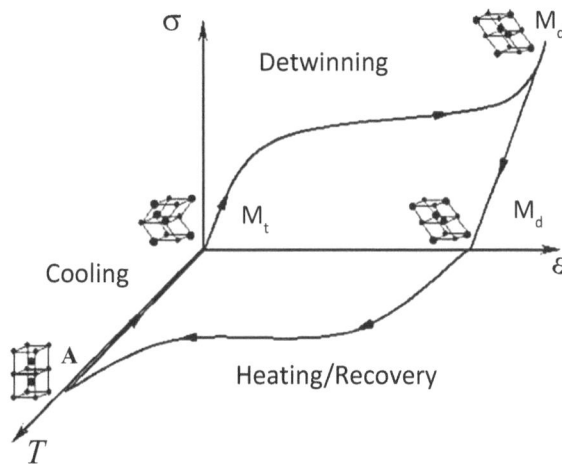

Fig. (5). Typical stress-strain-temperature curve of shape memory alloys.

Under austenitic transformation initiating temperature, self-accommodated martensite loads, the structure reorientates to perform self-accommodated strain *via* the twin boundaries movement and the detwinning of twin variants. Since the most appropriate variants are usually reoriented at very small load conditions (usually below 50 MPa) and it is difficult to measure the true elastic modulus of the material during loading, therefore usually measured during unloading process. For the applied load, the variants with the high orientation factor can reorient in relatively low stresses condition, and the variants with relatively unfavorable orientation factors require higher stresses to trigger the reorientation. The negative orientation variant gradient produces a linear region on the stress-strain curve during the loading process, followed by a 5-7% strain change with a little stress increase on the platform. Due to complete loading, every martensitic variant could only have chance to reorient, limiting the total strain increase.

After the most of the positive orientation martensitic variants have finished curve arrangement, stress-strain hardening curves happens. It is generally considered that this is an elastic deformation stage of martensite for detwinning and reorientating, followed by plastic deformation. Each martensitic variants' reorientation follows the law of conservation of energy. The reoriented driving force is derived from the applied stress, but the driving force to reform the twin martensite cannot be provided during the unloading process. Therefore, it can keep the shape of deformation after unloading and only recover the elastic strain in theory. But in fact, not all self-accommodated martensite can reorientate during the loading process, especially the unfavorable martensite variants. This contributes to localized stress inhomogeneity, which eventually becomes the driving force of the reorientation during the unloading process. Incomplete recovered strain after unloading can be recovered due to phase transition. The residual strain can be restored by heating to the austenitic transformation initiating temperature and this process ends when the temperature is higher than the austenitic transformation finished temperature.

(2) Superelastic Performance

The Ni-rich near-equiatomic Ti-Ni alloy appears austenite phase at room temperature. NiTi alloy has superelastic performance at room temperature, as explained in Fig. (**6**). At room temperature, the stress applied to the NiTi alloy induces martensite transformation. In stage (1) of low load, the austenite phase performs elastic deformation, producing a linear elastic strain. At the second stage (stage (2)), as the load increases, once the stress reaching the driving force required for the phase transition, the martensitic nucleuses inside the austenite grains. For the specific stress applied, the martensite nucleates in the orientation of the most favorable variants. During the third stage (stage (3)), as the load is continuing to increase, martensites grow and extend the entire austenite grains, and end at the grain boundaries. After stage (3), stress-induced martensite have an elastic deformation.

Above the austenite transition finished temperature, martensite phase is unstable. Thus, once unloading, the martensite phase will reverse to austenite phase. A linear elastic strain recovery occurs for the first unloading of martensite in stage (4). As unloading the phase transition platform, the reversal recover the strain. Thereafter, martensite converts into austenite, and austenite recovers the linear elastic strain at stage (5). If there is no plastic deformation during loading, the strain can recover completely during unloading, forming a superelastic ring and no residual strain existed in stage (6).

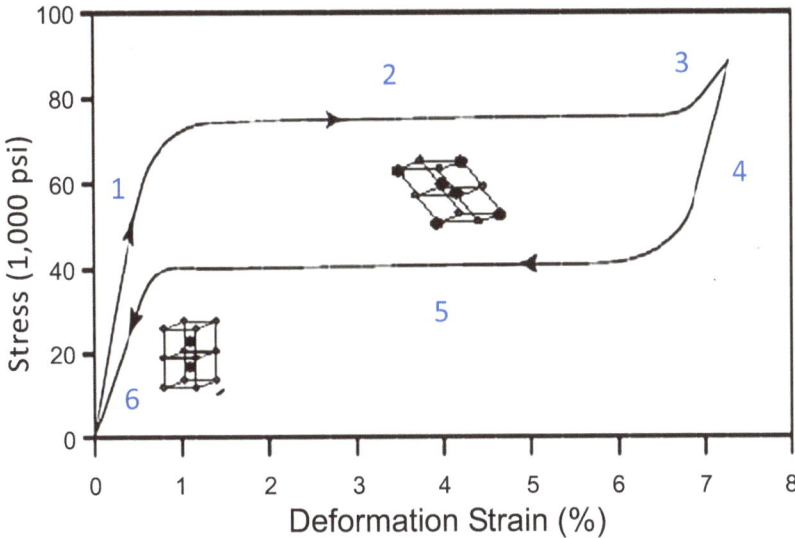

Fig. (6). Schematic diagram of typical superelastic stress-strain curve.

(3) Biocompatibility

As we all know, Ti has excellent biocompatibility and corrosion resistance [18]. On the other hand, Ni is toxic to the human body [19]. The corrosion resistance of NiTi alloy is between the Ti and the stainless steel, depending on the surface treatment and texture [20]. The researches of NiTi plastic arch wires demonstrated that the Ni ion release of the NiTi alloy martensite phase is lower compared to the NiTi alloy austenite phase [21]. Studies have demonstrated that NiTi alloy austenite could produce stress-reduced martensite by surface treatments to reduce the release of Ni ions [22]. It has been discovered that the NiTi alloy heat treatment and surface treatment (such as ion implantation) have been widely utilized to enhance the corresponding corrosion resistance [23]. The electrochemical polishing could also improve the corrosion resistance of the NiTi alloy and limit the Ni ion release level close to the 316 stainless steel [24, 25].

The NiTi alloy also has excellent biocompatibility except the good corrosion resistance [26]. Many long-term experiments *in vivo* have demonstrated that the host cells do not adverse reactions to the NiTi implants [27, 28]. A one-year research demonstrated that the NiTi and Ti particles implanted in the dura mater produced the same inflammation [29]. The porous NiTi alloys have the ability to fuse into surrounding cells similarly to Ti alloys, stainless steels and chromium cobalt alloys [30]. The bulk of the fusion is attributed to the TiO layer formed on the surface of the NiTi alloy. The TiO could not only act as an obstacle, preventing the release of Ni ions, but also displays biological activity [31]. Bone-like apatite layers are easily formed on the surface of TiO, induced by Ti-OH groups with similar crystal structures. In another reference [32], it was discovered that NiTi alloys could display biological activity generating anodic polarizations in virtue of H_2SO_4 and obtain different surface roughness values. This could increase the hydrophilicity of the surface, indicating that that the anodic oxidation had a higher surface energy that could promote the cell attachment and growth of the cultured humoral cells in surface cultures.

It is well known that the Nb has good biocompatibility. In the previous study [33], the niobium was implanted in the bone marrow of a soft tissue and in the femur for 2-4 weeks without inflammation, whereas no niobium was dissolved in the hard or soft tissues. Although a literature on the biocompatibility of the NiTi-Nb eutectic could not be directly found, in a certain research [34] the corrosion behavior of NiTi and Nb laser alloying was studied, where the NiTiNb eutectic formed on the surface. The corrosion resistance was measured in a Hank's solution, where both the NiTi and Nb were alloyed, the passive current density decreased and the pitting potential increased. This was a good indication of the biocompatibility of the NiTi-Nb eutectics [35].

NiTi alloy Biomedical Applications

Buehler *et al.* discovered the shape memory effect of a Ti-Ni alloy in 1962 and suggested that this material could be applied to dental implants. A few years later, Andreasen made the first superplastic support with the TiNi alloy. Since the shape memory alloy is applied in the minimally invasive surgery, the application of shape memory alloys in the field of biomedicine has been highly improved [36]. Now, certain complex medical treatments and surgical procedures are required to make use of certain precise and reliable micro-instruments for precise positioning to be achieved. This provides a golden opportunity for the commercial applications of shape memory alloys in the field of medicine. The applications of SMAs in medical devices are mainly divided into the following categories, such as orthopedics, neurology, cardiac medicine, interventional radiology and so on [37].

The superelastic behavior of SMAs is consistent with the stress-strain curves of the human bones and tendons, consequently the SMAs can be used in scaffolds within the human body. Kim *et al.* used the coil spring of the TiNi shape memory alloy in precision machining to form muscle fibers, which was mainly based on the principle of bionics, through the study of body movement process, creating a mesh worm model. More importantly, the TiNi shape memory alloy has a lot of good properties, such as good flexibility, high energy density, excellent flexibility and scalability. Stirling *et al.* executed a similar work, mainly studying the SMAs applications to the correction of the knee. Also, the shape memory alloy stents are in line with the curved vascular cavities, whereas the stainless steel stents often force the blood vessels to straighten. In addition, as an ideal scaffold, the superelastic behavior of SMAs could resist the physical shock in a normal physiological process, and also exert a restoring force outwards [38]. In 1983, Dotter's team firstly made the shape memory alloy stent, and consequently the amount of shape memory alloy stents rapidly increased in the world. Moreover, almost half are processed by the shape memory alloy. Furthermore, it also expanded the application of the shape memory alloy in other parts of the body. Currently, due to the necessity for minimally invasive surgery, the catheter-based surgery becomes increasingly common in order for the surgery to be less harmful. The application of shape memory materials to the localization accuracy of catheters at large angles of curvature has opened a new situation in both diagnosis and treatment [39, 40].

In recent years, with the wider utilization of shape memory alloys in biomedicine, an increased number of researchers have begun to pay attention to the fatigue and fracture behavior of shape memory alloys, due to the working life of SMAs having a big relationship with the corresponding fatigue and fracture properties. Therefore, researches of the fatigue and fracture behavior of SMAs are conducive to the corresponding application in the medical field.

Research About NiTi / NiTi-Nb Alloys

Grummon utilized a very efficient method of NiTi shape memory alloys connection based on the use of niobium as a melting point inhibitor, which could induce contact melting with NiTi at a temperature of 140 K below the melting point of NiTi [41, 42]. The joint portion of the NiTi shape memory alloy connections without the flux utilization is significantly firm and has good ductility, corrosion resistance, biocompatibility and could be processed under conventional industrial vacuum conditions. The discovery might have a profound effect on the use of NiTi shape memory alloys in complex aerospace structures and extend the range of NiTi shape memory applications.

As commonly known, the Ti-Ni-Nb system is the basis for wide-hysteresis shape memory alloys, mainly utilized in the space coupling device. The solubility of niobium in NiTi alloys is limited, such as in $Ti_{44}Ni_{47}Nb_9$ alloys, which exist in the form of spherical particles in the NiTi alloy matrices. The pure mechanical properties of these particles reduce the martensite starting temperature (Ms). Compared to the austenite starting (As), the corresponding temperature range expands.

The NiTiNb alloys with relatively high Nb content could solidify to eutectic structures [43]. Researches [44, 45] have demonstrated that pseudo-binary eutectic isotherms exist between the TiNi intermetallic compound and the pure Nb. The eutectic isotherm is lower than the isothermal melting temperature of NiTi alloys (140K), and the eutectic chemical composition is located near the $Ni_{38}Ti_{36}Nb_{26}$. Subsequently, when pure Nb contacts with NiTi at high temperature, spontaneous contact melting occurs, and only two phases appear during solidification: NiTi austenite and bcc-niobium. Titanium in the eutectic liquid has a high activity and it is easy to wet the NiTi surface, as well as many different materials, including the Al_2O_3. It is easy to flow into the capillary space and fully dissolve the surface oxide, as the flux utilization is avoided. It is critical to achieve good mechanical properties that no brittle ternary intermetallic phase exists in the cured joint structure.

The eutectic structure formed by solidification of this method has higher strength and good toughness. In addition, the complete shape memory performance and superelastic properties could be achieved at the joint by appropriate heat treatment. Therefore, this technology should be made to work out a variety of new designs to make use of the special performance of the NiTi alloys in the self-accommodated structure. As an example, it is possible to develop the extremely low density of the superelastic and shape memory NiTi alloys, such as the honeycomb and space frames, as well as shape memory/superelastic hybrid structures and new reactive composite reinforcements. It is also important that different NiTi alloy assemblies could be fused more easily. As an example, the superelastic component could be bonded to the NiTi casting or the NiTi foam materials incorporated into the shape memory device.

This connection process could be observed as a variant of the transient liquid phase (TLP) connection technique, which has been utilized in other alloy systems. However, in this method, the heating was stopped prior to the isothermal solidification occurrence and the liquid solidifies, forming a eutectic structure. Although, the authors have the ability to fabricate the real TLP connectors, the slow diffusion of niobium in the solid state makes the niobium unattractive for the conventional TLP processes, unless very thin films are used.

The connecting part had excellent mechanical properties. The tensile strength of the joint part is close to 0.8GPa and the fracture interface appeared to display microvoid coalescence and complete toughness. Microstructural studies have demonstrated the main factors controlling the evolution of microstructures (and oxide behavior) during the binding process.

Based on the current understanding of Ni-Ti, Ni-Nb and Ti-Nb systems and the reference data, the isothermal sections of Ni-Ti-Nb ternary phase diagrams at 20 °C and 900 °C were constructed, as presented in Fig. (7). The five ternary intermediate phases have been identified, all of which were located in the Ni-rich region of the diagram. More importantly, this region does not contain the known ternary intermediate phases in the Ti-rich side, partly due to the fact that both Ti and Nb were completely dissolved beyond the titanium α-β transition temperature (882 °C). The red lines connecting the NiTi and Nb on the 900 °C diagram were all possible trajectories of the liquid formed by the contact melting between the equiatomic NiTi and the pure Nb (assuming that solid-state diffusion was negligible).

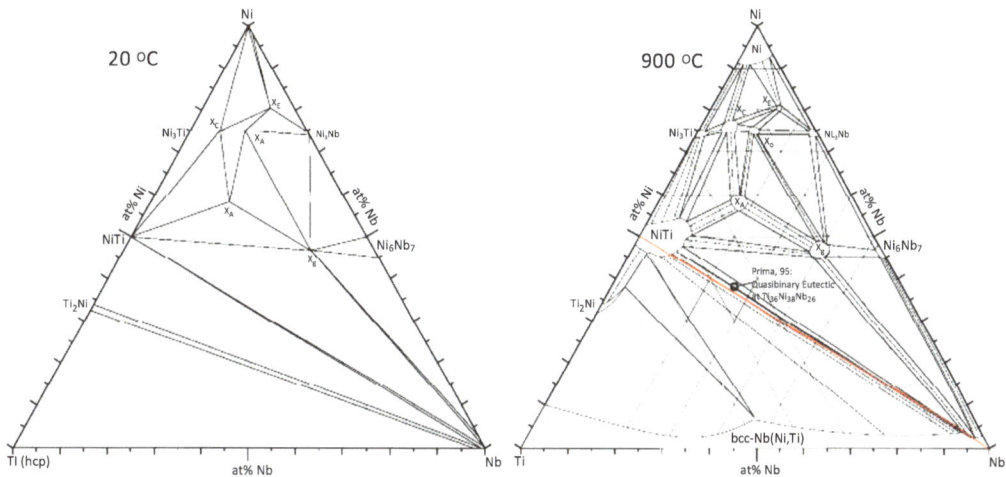

Fig. (7). 20 °C and 900 °C isothermals for Ni-Ti-Nb ternary system [41, 42, 44].

Fig. (8) schematically presents the basic characteristics of all Ni-Ti-Nb systems near the liquidus, with the three binary systems and liquidus surfaces on three end faces, and possible solutions of four phases with the liquid involved. In the presented five reaction planes, the U2 and U3 Class II quadrature balance were of particular importance. These two reaction isotherms include the liquid phase, the β-NiTi and the bcc-niobium. The fourth phase in the U2 was the compound Ni_6Nb_7, whereas for the U3, the Ti_2Ni phase completed the reaction plane. Therefore, between the U2 and the U3 (in the component space) was a large area of NiTi + Nb equilibrium, which indicated the existence of pseudo binary eutectic systems involving these terminal phases.

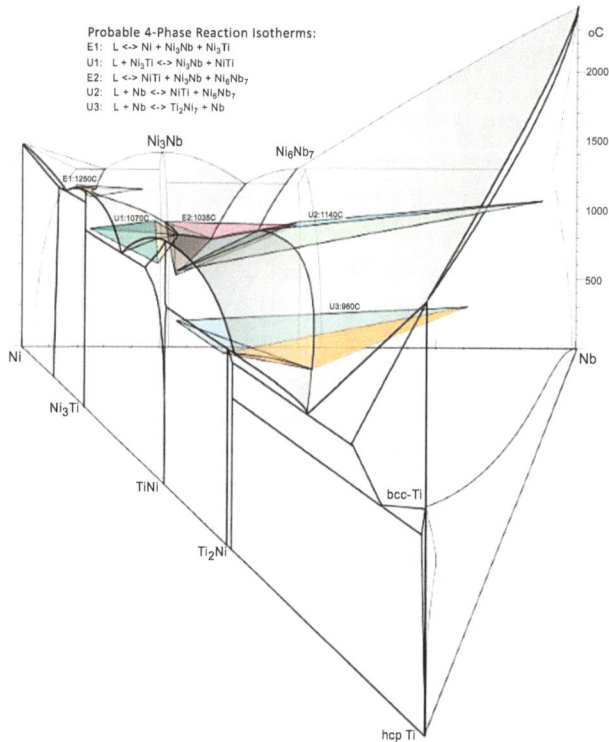

Probable 4-Phase Reaction Isotherms:
E1:   L <-> Ni + Ni$_3$Nb + Ni$_3$Ti
U1:   L + Ni$_3$Ti <-> Ni$_3$Nb + NiTi
E2:   L <-> NiTi + Ni$_3$Nb + Ni$_6$Nb$_7$
U2:   L + Nb <-> NiTi + Ni$_6$Nb$_7$
U3:   L + Nb <-> Ti$_2$Ni$_7$ + Nb

Fig. (8). Schematic ternary phase diagram for Ni-Ti-Nb system [41, 42, 44].

The pseudo-binary eutectic isotherms coincided with the maximum temperature of the two leaves of the three-phase NiTi + Nb + Liq region, which were created by the U2 and U3, as presented in Fig. (9). In this case, the ternary pattern was cut along the line connecting the NiTi and Nb to display the quasi-isotopic eutectic equivalent. The foreground diagram demonstrated that the NiTi + Nb + L3 phase field increased from the U3 at 960 °C and was pinned to an isothermal maximum at 1170 °C, consequently expelled again at 1140°C prior to the drop of U2.

The best estimation of the isentropic characteristics of the binary NiTi-Nb were presented, which could actually be a simple eutectic system. Although the microstructural studies in this work do not demonstrate the phases of the cured product, besides the B2-NiTi and bcc-Nb, it is indicated that the quasi-binary lines (track of the same concentration of Ni and Ti) overlapping the Ni$_6$Nb$_7$ + Nb three-phase field (extending downward from U2), approached 50 at % of Nb at 900°C. In this case, the NiTiNb system did not contain a "perfect" quasi-quaternary eutectic, whereas it contained along the suggested and more complex directions in the figure. In these results, it was possible that the simultaneous precipitation of a few NiTi and Nb worked as the pre-eutectic phase, whereas otherwise it would result in a curing, in a manner very similar to a completely quasi-binary eutectic system.

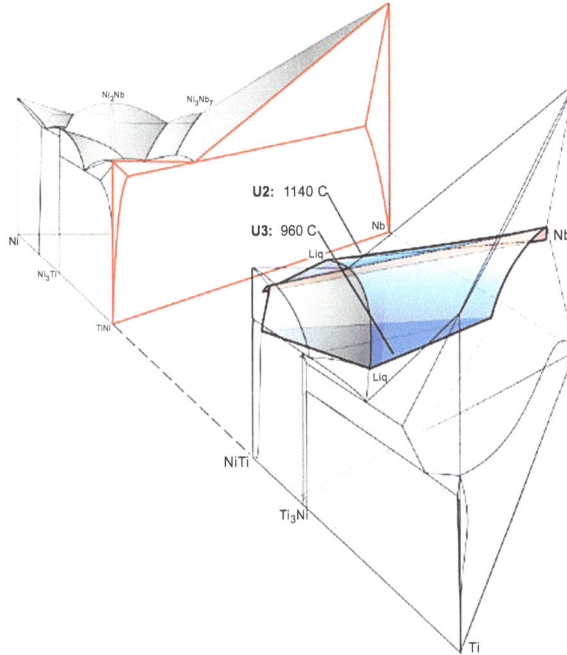

Fig. (9). Ni-Ti-Nb system cut along line junction of NiTi and Nb [41, 42, 44].

The process of contact melting could be explained by the "perfect" eutectic diagram in the figure. The red line in this case, represented the typical processing temperature (1180 °C) of the NiTi connection. Ignoring the presence of surface oxides and taking into account the monolayers (along with the associated average surface atomic density) on either side of the NiTi-Nb interface, it was discovered that the local interface composition was approximately 30 at% Nb. At 1180 °C, the solid-solid interface was thermodynamically unstable and sustained contact melting. If the solid diffusion was negligible during the liquid formation (a reasonable assumption for a short time, when the S/L boundary moved), the liquid composition must be located on the line connecting the pure NiTi and Nb ternary diagrams.

Once the liquid phase was formed, it was necessary to maintain the S/L interface balance with the two solids (NiTi and Nb) simultaneously, one on each side of the liquefaction zone. This meant that the liquid composition in contact with Nb (C2 in Fig. **4a**) was rich in Nb, which was richer than the Ni-Ti solid-liquid interface (C1). Although the average liquid composition must be between C1 and C2, these different interface compositions built a gradient in the liquid that drove the cross diffusion of NiTi and Nb between the two interfaces. As an example, when niobium diffuses through the liquid to the NiTi interface, the local equilibrium is destroyed and can only be reconstructed by the appropriate amount melting of NiTi and vice versa. In this way, continued to melt (approximately 2.2 volumes consumed per volume Nb) until the niobium was depleted. At this time, the liquid composition must move rapidly to the intersection of the isotherm at the processing temperature and the liquidus on the NiTi (C1). Subsequently, the interface balance could be built and the melting stopped. (As in the TLP process, the solid diffusion of Nb into the NiTi section would subsequently result in a slow isothermal formation of the constituent eutectic NiTi phase, whereas it has been observed that the method is very slow at the temperatures used in these experiments) Since the liquid composition (C1) was rich in NiTi, compared to the eutectic composition, the first eutectic NiTi phase was firstly cured, followed by a eutectic solid formation at 1170 °C as indicated in Fig. (10) about the quasibinary NiTi-Nb phase diagram.

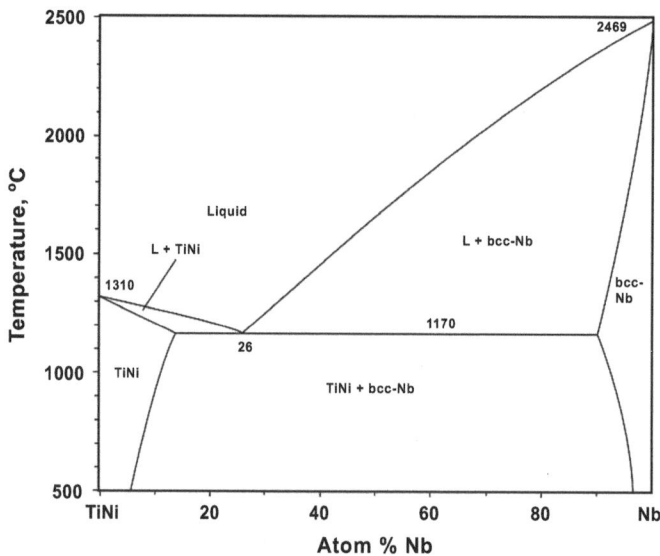

Fig. (10). Quasibinary NiTi-Nb phase diagram [41].

What has just been described is consistent with the time-resolved observation (presented in Fig. **11**) of the microstructure of the NiTi-Nb-NiTi pair maintained from 1 to 120 seconds at 1185 °C and followed by furnace cooling at various times. In the first second, the melting was significantly executed, whereas the microstructure contained a high amount of unreacted niobium, eutectic NiTi + Nb and the raw material NiTi solidification zone. As the brazing duration increased, the niobium was stably consumed and depleted at approximately 30 to 40 seconds at a reduced rate under the aforementioned temperature.

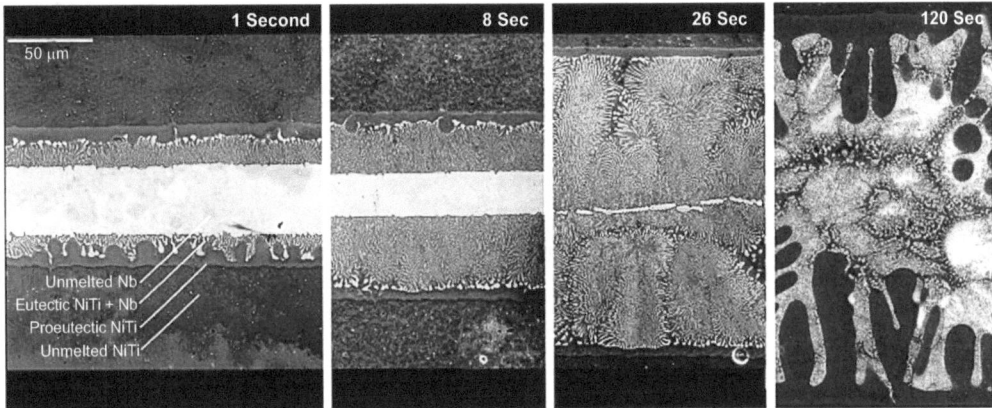

Fig. (11). Time-resolved development of brazing microstructure [45].

The melting rate was limited by the diffusion coefficient in the liquid and the concentration gradient established by the equilibrium preservation at two different S/L interfaces simultaneously. In the experiment, it was estimated that the liquid phase diffusion coefficient (approximately 10^{-9} m^2/s) was approximately the same as the melting rate observed at 1185 °C. The decreased rate of liquid formation as time was consistent with the nominal diffusion coefficient, whereas it was combined with the increased path length for diffusion, and the latter produced more liquid forms. The net effect of the liquid fraction increase was to flatten the concentration gradient. As presented in Fig. (11), the relative amount of primary NiTi was reduced as the niobium was consumed. This situation might occur because the liquid composition was a function of the melting rate. Initially, when the rate was high, the liquid could be relatively enriched with NiTi, whereas as the rate slowed down, the niobium content might increase because the melting dynamics on the Nb side were inherently slower and coincided with the high melting point of Nb.

The initial test demonstrated that the strength and ductility of the brazed joint was quite good. Fig. (12) presents a typical stress-strain curve of a 3 mm thick hyperelastic NiTi alloy with the butt joint at the center of the tensile specimen.

The samples were brazed with pure niobium at 1180°C for 6 minutes, following furnace cooling and finally annealed at 350°C for 90 minutes prior to testing. The fractures occurred at just below 800 MPa and the platform associated with the stress-assisted martensite formation was clearly observed. The dotted line in Fig. (12) represents the typical superelastic behavior observed subsequently to several vibratory cycles in the docked joint sample, indicating that the excellent superelasticity could be achieved by the heat treatment following the proper brazing. The illustrations in the figure demonstrated the typical broken concave cup-conical fracture surfaces of the eutectic microstructures of the brazed joints, indicating that the brazing material was completely extensible [41, 42, 44].

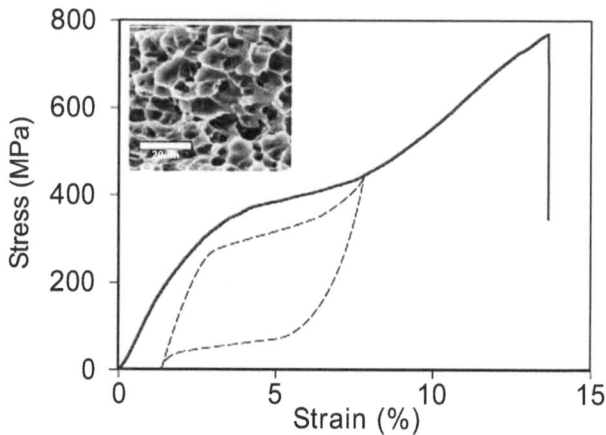

Fig. (12). Stress - strain curve for NiTi-Nb joint.

Wang [46 - 48] utilized the Nb powder *in situ* reaction to connect the NiTi wires,and studied the microstructure, the microhardness and the toughness of the eutectic phase. In addition, the NiTi/Nb interface was also observed.

Meaning and Content of This Research Work

The NiTi alloy has been one of the most considered engineering materials due to the unique properties, such as superelasticity and shape memory effect. The NiTi alloy has good biocompatibility, excellent corrosion resistance, low elastic modulus, good ductility and other mechanical properties [49 - 51]. It is widely utilized in many fields such as the biomedical, the aerospace, intelligent materials and so on. The bulk NiTi alloy with a pore structure is called the porous NiTi alloy. On one hand, the elastic modulus could be reduced to constitute it similar to the elastic modulus of the bone tissue, avoiding the stress shielding effect due to the NiTi alloy not matching the bones' elastic properties as a bone graft. On the other hand, it facilitates the growth of bone tissues into the open pore structure, consequently enabling the bone tissue to closely connect with the NiTi graft [52]. In addition, due to the NiTi alloy's good biocompatibility and excellent corrosion resistance, the porous NiTi alloy has become one of the best choices for biomedical materials. The porous NiTi alloy is prepared by the powder metallurgy technology, due to the corresponding high melting point. NaCl particles are utilized as pore forming agents, mainly including self-propagating high temperature sintering (SHS), plasma sintering (SPS), hot isostatic pressing (hip) and so on [53 - 55]. In contrast, most porous NiTi alloys prepared by this method belong to the foam materials. The pore structure is arranged randomly. The pore size and foam porosity can be controlled only by the size and content control of the pore forming agents and massive closed pores exist, which are not conducive to the bone tissues growing into the material interior. The NiTi lattice material can be flexibly adjusted according to the practical application conditions, due to the corresponding porous structure, and the pore structure is interconnected to assist the bone tissue growth, promoting the growth itself.

The project on the microstructure and mechanical properties of the interface between the Nb foil and the NiTi alloy contributes in the advantage comprehension of the Nb utilization as a NiTi alloy connection material, which provides theoretical support for the preparation of the NiTi lattice material and promotes the significantly extensive NiTi alloy application.

PREPARATION AND ANALYSIS OF MATERIALS

Grummon *et al.* demonstrated that the addition of the third element (Nb) was utilized to decrease the NiTi melting point, in order to produce the liquid *via* Nb contact with the NiTi alloy at high temperature [41, 42]. Subsequently cooling down to the room temperature, the NiTi wire could form a good connection, and the joints were firmly connected, displaying good ductility, corrosion resistance and biocompatibility. This could lead in a series of problems avoidance in traditional welding methods, such as the brittle phase in the joints, where the heat affected area is quite high-sized, resulting in joints of poor performance. Therefore, in this study, the NiTi/ NiTi-Nb lattice materials were prepared by the Nb foil utilization as material for the NiTi wires connection.

In this experiment, the Nb foil and NiTi NiTi wires utilization produced the NiTi-Nb eutectic connecting the NiTi wires, thereby the NiTi-Nb lattice materials in variable paraments were prepared, the microstructure of the boundary joints of the NiTi/ NiTi-Nb lattice materials was characterized and the phase transition and elastic properties at the connected area were measured.

Materials Preparation

The NiTi/NiTi-Nb lattice materials were prepared by the NiTi wire and Nb foil sintering in an argon atmosphere high temperature furnace, followed by aging treatment. The schematic diagram of the fabrication processing of NiTi/NiTi-Nb lattice-structured material is presented in Fig. (**13**). The high temperature sintering furnace filled with Ar gas is presented in Fig. (**14**).

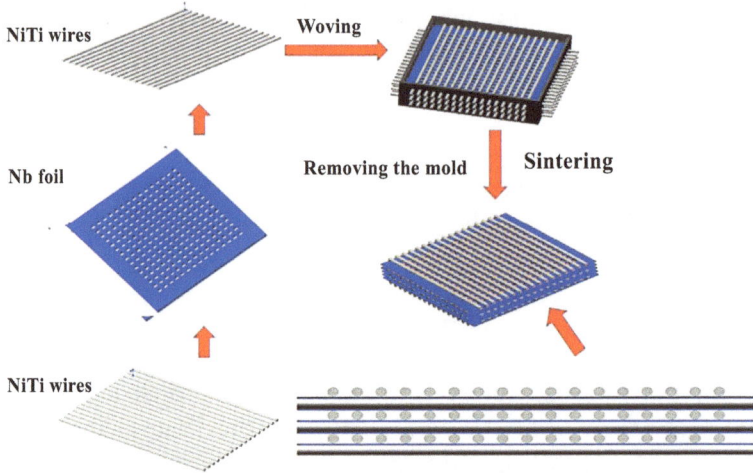

Fig. (13). Schematic diagram of fabrication processing of NiTi/ NiTi-Nb lattice-structured material.

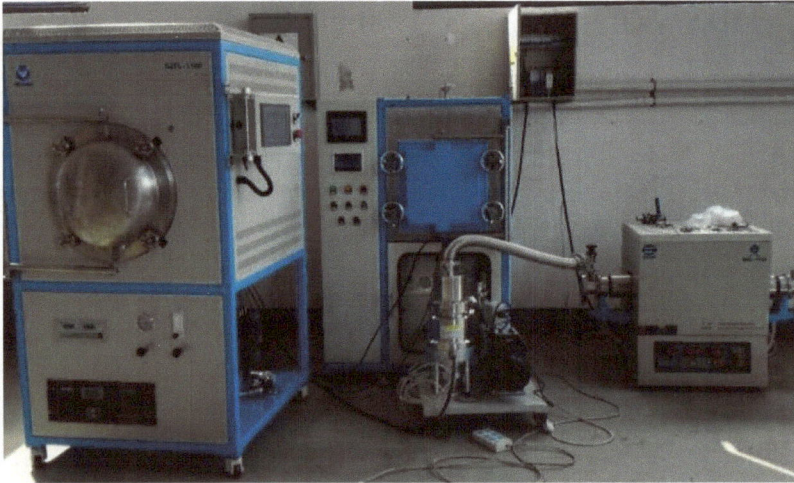

Fig. (14). High temperature sintering furnace filled with Ar gas.

Experimental Materials

The experimental materials used in this study included the NiTi wire, the Nb foil and the stainless steel frame. Among these, the NiTi wire grades were Ti-50.9 wt% Ni, the diameter was 1mm, as purchased from the Jiangyin Farersheng Pyle New Materials Technology Co.Ltd, whereas the specific chemical composition is presented in Table **2**. The purity of the Nb foil was 99.95% and the corresponding thickness was 0.5 mm. The stainless steel frame length, the width and the height were 40mm, 40mm and 10mm, and the thickness was 1mm.

Table 2. Chemical composition of experimental NiTi wires (mass %).

Ni	Ti	O	C	N
55.78	44.17	0.03	0.01	0.01

Frame Design

The frame utilized in this experiment was a stainless steel square tube. The corresponding length, width and height were 40mm, 40mm and 10mm, respectively, and the thickness was 1mm. Using the WEDM cut the frame as presented in Fig. (**15**). Among these dimensions, the groove width was 1mm, the height was 8mm and the groove center spacing was 2mm. The groove on both sides of the center line was evenly distributed, whereas opposite the two sides of the groove was the facing.

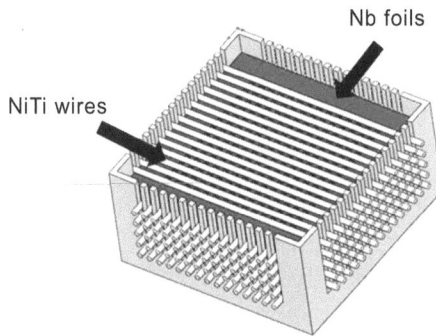

Fig. (15). Frame schematic diagram of NiTi/NiTi-Nb lattice-structured material.

Sintering Process

In order to make the NiTi wire connection, the Nb foil was used as the melting point inhibitor, sintered at high temperature to produce liquid wetting surface, forming a strong metallurgical bond subsequently to cooling.

The NiTi of 50 mm in length and of 1 mm in diameter along with a rectangular Nb foil of 38 mm in both length and width were placed in an acid solution (HF: HNO_3: H_2O = 1: 3: 10 (volume ratio)) for pickling, where the oxide layer of the NiTi wires and Nb foils were removed, in order to be in contact with each other completely, promoting sintering. Subsequently, the NiTi wires and the Nb foils were laid out as presented in Fig. (**16**), and the samples were wrapped with Ti foils to avoid contact with oxygen and consequently placed in a high temperature sintering furnace filled with Ar gas. With reference to the Grummon's sintering process [41, 42], the specific sintering process is presented in Fig. (16).

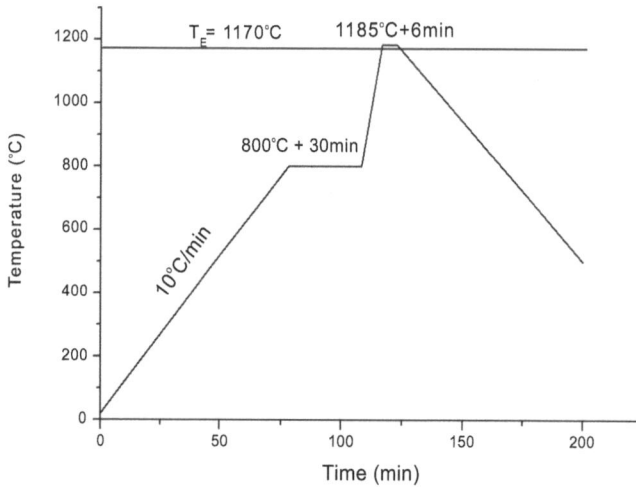

Fig. (16). Sintering processing curve of NiTi/NiTi-Nb lattice-structured material.

Analysis Test Methods

In this paper, the composition, the phase distribution, the microstructure distribution of the interface junction of the NiTi wires and the Nb foils in the NiTi/NiTi-Nb alloy were mainly characterized as well as the mechanical properties of the connection area, such as the elastic modulus and the deformation recovery rate. The microscopy methods for the characterization and analysis included: the optical microscope (OM); the energy dispersive spectroscopy (EDS) for surface scanning, line scanning and spot scanning to determine the elements distribution and phase composition at the interface; nanoindentation test for the determination and characterization of the elastic modulus and deformation recovery ability.

Metallographic Microscope

All metallographic images of this subject were taken with the ZEISS AXIO Imager A1m optical microscope. The specific metallographic sample preparation process in this experiment was as follows: The sintered samples were cut by WEDM and the solid parts were utilized to prepare metallographic samples. The size of metallographic samples was 0.5mm*2mm*2mm. The sintered specimen was embedded into the epoxy resin by a heat insulator. All 80 #, 240 #, 400 #, 600 #, 1200 #, 1200 #, 1500 # and 2000 # metallographic sandpapers were utilized in the automatic grinding machine, respectively; the samples were polished with the polishing solution of Cr_2O_3 until no scratch could be observed under the metallographic microscope; the sample surface was cleaned by anhydrous ethanol and finally a hair dryer was utilized to dry the sample to avoid oxidation. The NiTi/NiTi-Nb alloy subsequently to polishing was etched by the Kroll solution to make the interface between the grain boundary and the phase more visible. The specific ratio of the Kroll solution was HF: HNO_3: H_2O = 1: 3: 10 (volume ratio). A cotton ball dipped in the Kroll corrosive liquid by the tweezers smear was used on the NiTi/NiTi-Nb alloy sample surface uniformly; a significant amount of water was utilized to clean the sample surface following 10s of corrosion, and finally a moderate amount of anhydrous ethanol was sprayed on the sample surface. The surface was dried with a hair dryer.

The corroded NiTi/NiTi-Nb alloy was photographed under the ZEISS AXIO Imager A1m optical microscope.

Energy Dispersive Spectrometer

The energy dispersive spectrometer utilized in this subject was an accessory during the scanning electron microscopy. The energy dispersive spectrometer mainly utilized the X-ray characteristic wavelengths of the respective elements to analyze the species and content of the micro-component elements of the material. It was utilized in conjunction with scanning electron microscopy and transmission electron microscopy. In this experiment, the energy dispersive spectrometer was utilized for surface scanning, line scanning and spot scanning to analyze the element distribution and phase element composition of the diffusion layer in the interface between the Nb foils and the NiTi wires in the NiTi/NiTi-Nb alloy.

Nanoindentation Test

The nanoindentation was mainly utilized to determine the microhardness and Young's modulus of the micro-nano-sized film materials and the results were obtained from the force curve and the penetration depth. The nanoindentation diagram is presented in Fig. (**17**). Figs. (**18** and **19**) display the relation curve of load and indentation depth in the nanoindentation test and the indentation diagrams of the specimen prior to and following compression. The hardness and elasticity modulus from testing can be calculated according to the following equations:

$$H = \frac{F_{max}}{A} \qquad\qquad (2)$$

$$E_r = \frac{1-v^2}{E} + \frac{1-v_i^2}{E_i} \tag{3}$$

$$S = \frac{dP}{dh} = \frac{2}{\sqrt{\pi}} E_r \sqrt{A} \tag{4}$$

where, F_{max} is the maximum load, A is the projected area of indentation, S is the slope of the upper part of the unloading curve, E_r is the valid elastic modulus, E is the elastic modulus of the material measured, V is the Poisson ratio of the measured material, E_i is the elastic modulus of the presser material and V_i is the Poisson ratio of the head material.

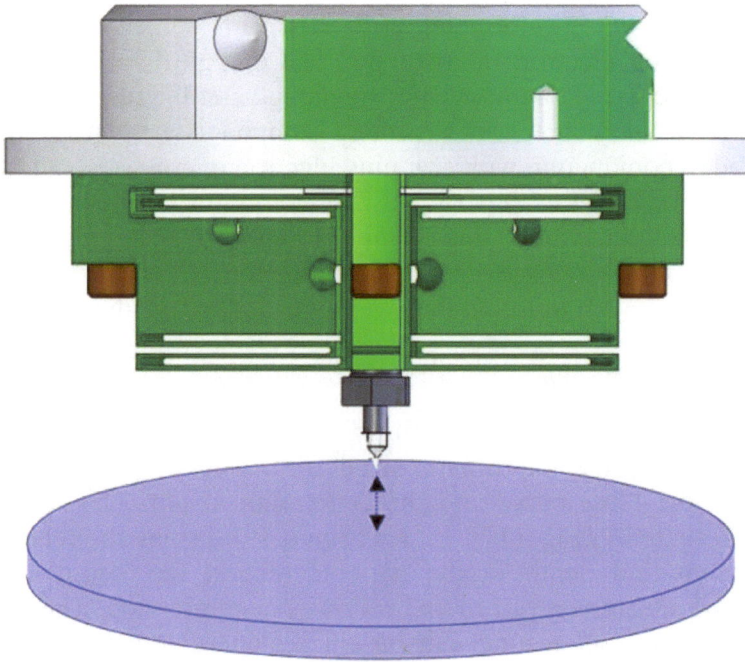

Fig. (17). Schematic diagram of nanoindentation instrument.

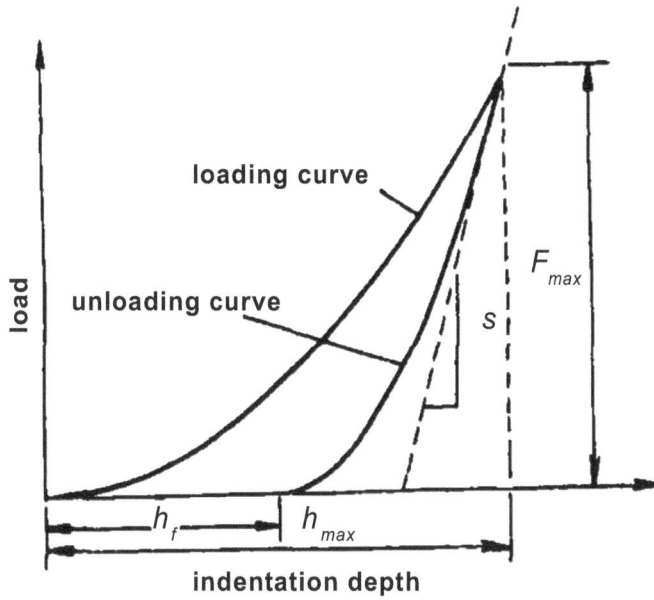

Fig. (18). Load-height curve in nanoindentation test.

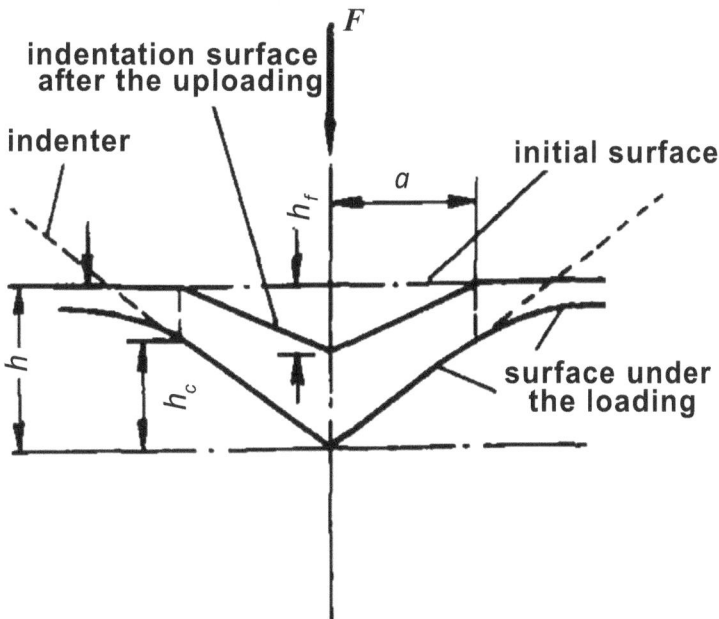

Fig. (19). Indenter schematic diagram prior to and following indentation tests.

MICROSTRUCTURE AND MECHANICAL PROPERTIES OF A SINTERED NITI/NITI-NB ALLOY

Experimental Results and Analysis

Characterization of Sintered Metallographic Microstructure

Fig. (20) presents a NiTi/NiTi-Nb alloy sample prepared by the Nb foils and NiTi wires sintering in high temperature sintering furnace filled with Ar gas at 1185 °C for 6 min. As presented from Fig. (20a), the NiTi wires formed a compact metallurgical bond with the Nb foils *via* the fusion zone between them. It could be observed from Fig. (20b) that the grain size of the NiTi matrix was very small and the diameter was approximately 1μm. A high number of fine pores existed on the surface, due to the diffusion of Ni and Ti atoms into the Nb foil. Fig. (20c) presents the microstructure micrograghs of interface between NiTi layers and the fusion zone, where the microstructure of the fusion zone was short or striped, and the orientation of the strip structure differed in different domains. As presented in Fig. (20d), the NiTi wires, the fusion zone and the Nb foils were all closely boned, and the metallurgical structure of the fusion zone was similar to the metallurgical structure of the fusion zone of Fig. (20c).

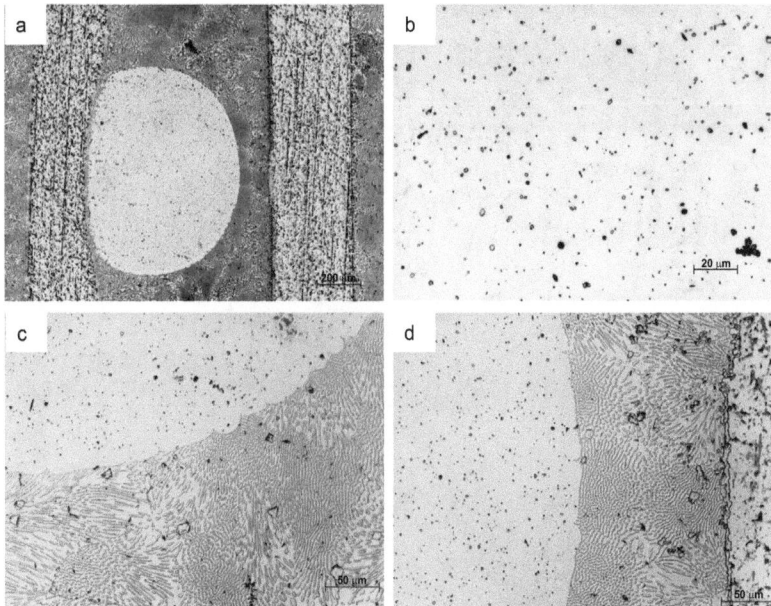

Fig. (20). Microstructure micrograghs of NiTi/NiTi-Nb alloys sintering at 1185 °C for 6min: (**a**) interface between NiTi wires and Nb foils; (**b**) NiTi matrix; (**c**) interface between NiTi wires and fusion zone; (**d**) interface between NiTi matrix, fusion zone and Nb foils.

Diffused Layer Scanning

Fig. (**21**) presents the NiTi wires and Nb foils heated to 1185 °C (close to NiTi-Nb pseudo-binary eutectic reaction temperature). As presented in Fig. (**21a**), the diffusion layer region included region I: Nb foil; region II: NiTi-Nb eutectic region; region III: primary NiTi region and region IV: NiTi matrix region. Once the NiTi-Nb pseudo-binary eutectic temperature (1170 °C) was reached, the eutectic liquid phase was produced first, and the chemical composition of the NiTi-Nb eutectic was similar to the $Ni_{40}Ti_{40}Nb_{20}$ composition, as referred in the literature [56, 57]. Subsequently, the liquid phase was required to maintain the balance of the Ni-Ti solid-liquid interface on both sides and the Nb solid-liquid interface, whereas apparently the Nb content of the solid-liquid interface side was higher compared to the NiTi solid-liquid interface. In the same way, the NiTi content of the NiTi solid-liquid interface was higher compared to the Nb solid-liquid interface, which led to the Nb and NiTi concentration gradient at the Nb-solid and NiTi solid-liquid interfaces, driving the cross diffusion of the NiTi and the Nb between the two solid-liquid interfaces. According to the NiTi-Nb binary pseudo-eutectic phase diagram, the NiTi content at the NiTi solid-liquid interface was higher compared to the eutectic. It could also be observed that the NiTi content was higher at the interface of NiTi compared to the eutectic, the primary NiTi phase precipitated during the cooling and the NiTi-Nb eutectic phase formed at 1170 °C subsequently. Similarly, the Nb content in the Nb-solid interface side was higher compared to the eutectic, the primary β-Nb (bcc) preferentially precipitated during cooling, and consequently the NiTi-Nb eutectic structure formed at 1170 °C. According to the results presented in Fig. (**21b**) measured with line scanning from left to right by the EDS device according to the position of the blue line in Fig. (**21a**), the Nb content fluctuated apparently near regions I and II. When the line scanning went through the light-colored area, the Nb content increased significantly, and the Ni and Ti content decreased correspondingly; accordingly, when the line scan went through the dark area, the Nb content decreased significantly, and the Ni and Ti contents increased significantly. On the one hand, the Nb content was high at the Nb solid-liquid interface and the primary Nb-rich phase formed during cooling; on the other hand, the Nb solution was low in the NiTi alloy, leading to the Nb separation from the parent phase of NiTi. In addition, it could be observed from Fig. (**21b**) that the Ti content fluctuated and the Ni content remained stable in the diffused dark region. It possibly occurred because the atomic size of Nb (r = 0.208 nm) was similar to atomic size of Ti (r = 0.200 nm), and differed highly from the atomic size of Ni (r = 0.162 nm). Also, the Ti and Nb could form infinite solid solutions, therefore the Nb atom replaced the Ti atom, which was consistent with the conclusion in the references [58]. In Fig. (**21b**), the Nb content at the interface between regions I and II decreased significantly, and the Nb content was basically stable in regions II and

III, which might be caused by the Nb foils and NiTi wires continuous consumption when the liquid phase was produced during the 6 min period over 1170 °C, whereas the Nb, Ni and Ti diffused across inside the liquid phase, making the chemical composition relatively uniform, subsequently forming a relatively uniform NiTi-Nb eutectic structure during cooling.

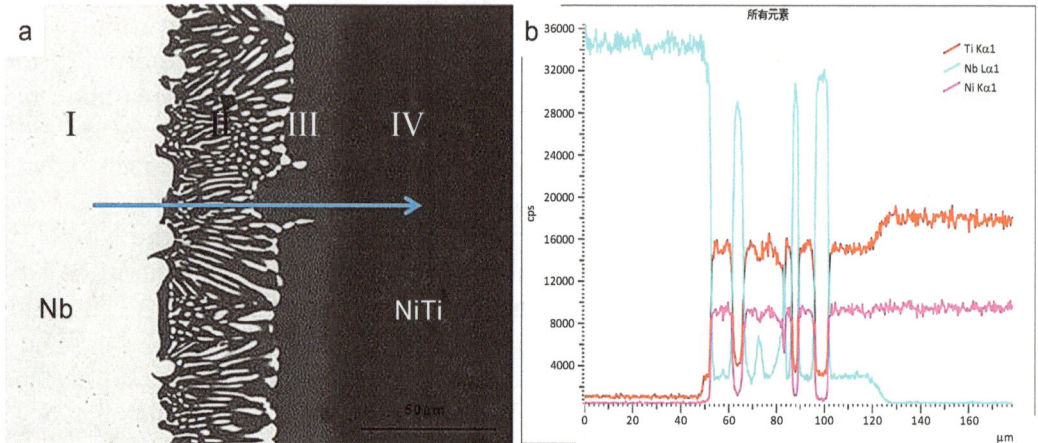

Fig. (21). Diffusion layers and line scanning results of polished surface of brazed joint between Nb foils and NiTi wires: (**a**) SEM micrographs of diffusion layers; (**b**) results of line-scanning of diffusion layers.

Microstructure Phase and Elemental Composition

Fig. (**22**) presents the SEM microstructure of the NiTi wires and Nb foils heated to 1185 °C (near NiTi-Nb pseudo-binary eutectic reaction temperature), whereas this temperature was maintained for 6 mins. It could be observed from Fig. (**22a**) that three essential phases existed in the region: the NiTi matrix, the Ti-rich phase of polygonal structure and the short rob-like and long striped Nb-rich phase. A high number of equiaxed and elongated eutectic structures (E) existed, consisting of the NiTi and Nb-rich phases at region II as well as the short rob-like and long striped Nb-rich phase (Nb') along with a low amount of multiple-face phase (Ti'), and a primary eutectic NiTi phase (PE) existed near region III. The chemical composition of each phase is presented in Table **3**, whereas the chemical compositions of the NiTi matrix, the NiTi phase, the Ti-rich phase, the Nb-rich phase and the eutectic structure were 51.18Ni-48.81Ti-0.01Nb, 50.26Ni-43.60Ti-5.84, 34.25Ni-58.26Ti-7.49Nb, 3.91Ni-15.33Ti-80.76Nb and 37.37Ni-39.27Ti-25.36Nb, respectively. Among these, the Nb-rich phases were globular or streaky, whereas the eutectic structure was mostly strip-like, and the grain orientation was different among the grains. Similarly to the microstructure of 40Ni-40Ti-20Nb [59], as grains A and B are presented in Fig. (**22a**), the Ti-rich phase had a polyhedral structure. As presented in Fig. (**22b**), a high number of the surface convex structures existed on the boundary between the NiTi-Nb eutectic structure and the NiTi matrix, which was martensite (M), whereas rare in the eutectic interior. The martensite initiation transition temperature (M_s) depends on the Ni/Ti [57] in the B2 matrix. As presented in Fig. (**22b**), the contents of Ni/Ti and Nb in regions II and III were both higher compared to the NiTi matrix, resulting in the amount of the martensite initiation decrease. At the boundary of the eutectic area, a coarse structure existed, which is called the mesophase. The mesophase had an Nb-rich phase precipitate with an equiaxed diameter of approximately 3 μm. Elongated precipitates existed near the tiny tissue with a diameter of approximately 8 μm. A few multiple-face precipitates were observed in the mesophase, as presented in the small black areas. The chemical composition of the precipitates was 58.26Ti -34.25Ni-7.49Nb as measured from the EDS point scanning, which is considered as Ti_2Ni, whereas certain scholars defined the precipitates as Ti_2 (Ni, Nb) compounds. The Ti_2Ni phase size varied and the corresponding diameter ranged between 2μm-10μm. The literatures demonstrated that: excluding the Nb content in the alloy, the Ti_2Ni was only formed when the Ni/ Ti was below 1 [60, 61]. The Ni/ Ti chemical composition of the NiTi wires utilized in this experiment was 0.84, and the Ti_2Ni phase could be observed in the eutectic region. The chemical composition of fine the eutectic region was 37.37Ni-39.27Ti-25.36Nb, which was similar to the published eutectic chemical composition Ti-40Ni-20Nb, as measured by the EDS. According to the pseudo binary NiTi-Nb eutectic phase diagram established by Grummon, the solid solubility of Nb in NiTi was in the range of 10-14 at % at the eutectic temperature, whereas a higher solid solubility was discovered in this study.

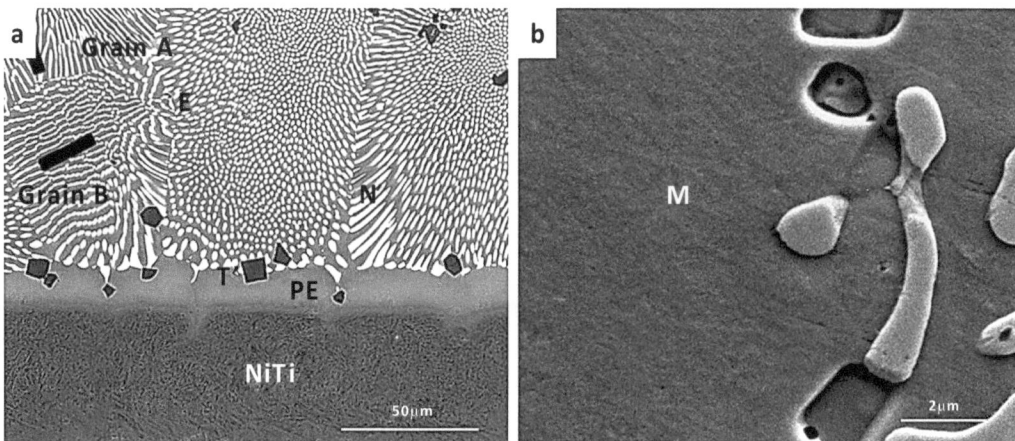

Fig. (22). SEM micrographs of polished surface of brazed joint between Nb foils and NiTi wires: (**a**) microstructure of brazing regions; (**b**) martensite phase in NiTi matrix and fusion zones.

Table 3. Results of EDS point scanning of fusion zone near NiTi wires.

Spectral line	Ni (at %)	Ti (at %)	Nb (at %)
NiTi	51.18	48.81	0.01
PE	50.26	43.60	5.84
T'	34.25	58.26	7.49
Nb'	3.91	15.33	80.76
E	37.37	39.27	25.36

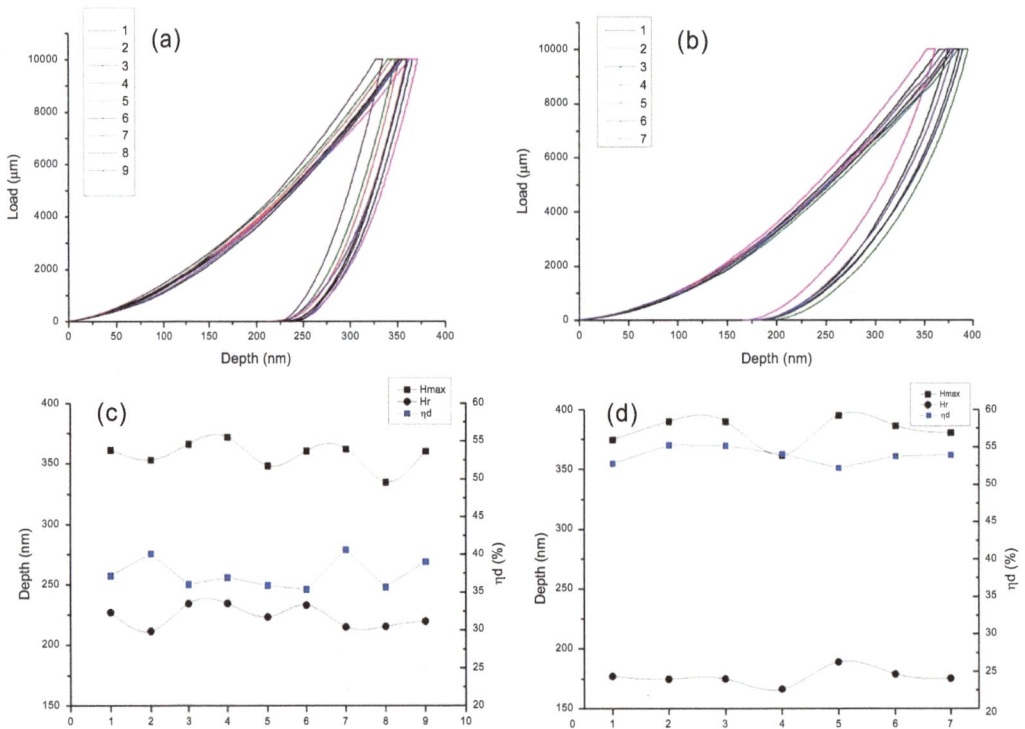

Fig. (23). Nanoindentation results of interface between Nb foils and NiTi wires in NiTi/NiTi-Nb alloys: **(a)** indentor load-depth curve in fusion zone; **(b)** indentor load-depth curve in NiTi matrix; **(c)** indentor depth prior to and following loading and deformation recovery rate curve in fusion zones; **(d)** indentor depth prior ti and following loading and deformation recovery rate curve in NiTi matrix.

Figs. (**23a-b**) presents the force-displacement curves of fusion zone 1 and NiTi matrix region 2, respectively. Figs. (**23c-d**) present the images of the maximum indents depth (h_{max}), the residual depth following unloading (h_r) and the strain recovery rate (η), according to the data as in Figs. (**23a-b**). As it could be observed in Fig. (**23c**), the maximum indentation depth of fusion zone 1 was 335-

350nm, and the residual depth following unloading (h_r) was 210-240 nm. The strain recovery rate can be calculated by the following equation:

$$\eta = \frac{h_{max} - h_r}{h_{max}} \times 100\% \tag{5}$$

The strain recovery rate (η) ranged between 35% -40%; Similarly, as presented in Fig. (**23d**), the maximum depth of nanoindentation (h_{max}) in the NiTi matrix region 2 ranged between 375-400 nm, whereas the residual depth (h_r) following unloading ranged between 170-200 nm and the strain recovery rate (η) ranged from 52% to 55%. The partial solid solution of Nb with the parent phase matrix contributed to the solid solution strengthening. In addition, the Nb-rich particles were dispersed in the eutectic region, contributing to the dispersion strengthening, enhancing the strength of the fusion range, making the maximum depth of the fusion zone lower compared to the NiTi matrix region. The soft β-Nb phase of the fusion zone released the elastic strain during the martensitic transformation *via* plastic deformation, contributing to the martensite reversal transition driving force decrease, preventing strain recovery. In addition, the higher content of Ni/Ti decreased the martensitic transformation temperature and inhibited the strain recovery effect, ultimately leading to the strain recovery of fusion zone 1 strain lower compared to the NiTi matrix 2.

CONCLUSIONS

The main conclusions of this subject were as follows:

1. NiTi wire and Nb foil were heated to 1185°C and this temperature was retained for 8 min in a high temperature sintering furnace filled with Ar gas, leading to the NiTi/NiTi-Nb alloy preparation. Between the NiTi wires and the Nb foils two distinct diffusion regions existed: the NiTi-Nb pseudo-eutectic region and the primary eutectic NiTi region.
2. The fusion zone between the NiTi wire and the Nb foil mainly consisted of a stripe-like Nb-rich phase and a NiTi-rich phase, and also a multiple-face structure of the Ti-rich phase was observed. The result of the point scanning demonstrated the chemical compositions of the NiTi matrix, the primary eutectic NiTi phase, the Ti-rich phase, the Nb-rich phase and the pseudo-eutectic structure as 51.18Ni-48.81Ti-0.01Nb, 50.26Ni-43.60Ti-5.84Nb, 34.25Ni-58.26Ti-7.49, 391Ni-15.33Ti-80.76Nb and 37.37Ni-39.27Ti-25.36Nb, respectively.
3. The strength of the fusion zone between the NiTi wire and the Nb foil was higher compared to the NiTi matrix.

CONFLICT OF INTEREST

The author (editor) declares no conflict of interest, financial or otherwise.

ACKNOWLEDGEMENTS

This work was supported by DARPA under the grant W91CRB1010004 and the National Science Foundation under grant No. 51302168 of China. Shanghai Pujiang Program:15PJD017. Medical Engineering Cross Research Foundation of Shanghai Jiao Tong University under Grant No. YG2014MS02 and SMC-ChenXing Project Shanghai Jiao Tong University. The authors acknowledge Prof. David C. Dunand from Northwestern University (USA) for useful discussions.

REFERENCES

[1]   Huang WM, Ding Z, Wang CC, Wei J, Zhao Y, Purnawali H. Shape memory materials. Mater Today 2010; 13: 54-61.
        [http://dx.doi.org/10.1016/S1369-7021(10)70128-0]

[2]   Ölander A. An electrochemical investigation of solid cadmium-gold alloys. J Am Chem Soc 1932; 54: 3819-33.
        [http://dx.doi.org/10.1021/ja01349a004]

[3]   Kurdjumov GV, Khandros LG. First reports of the thermoelastic behaviour of the martensitic phase of Au-Cd alloys. Dokl Akad Nauk SSSR 1949; 66: 211-3.

[4]   Greninger AB, Mooradian VG. Strain transformation in metastable beta copper-zinc and beta copper-tin alloys 1937.

[5]   Burkart MW, Read TA. Diffusionless phase change in the indium-thallium system. Trans Am Inst Min Metall Eng 1953; 197: 1516-24.

[6]   Sakamoto H, Otsuka K, Shimizu K. Rubber-like behavior in a Cu-Al-Ni alloy. Scr Metall 1977; 11: 607-11.
        [http://dx.doi.org/10.1016/0036-9748(77)90118-1]

[7]   Kauffman GB, Mayo I. The story of nitinol: the serendipitous discovery of the memory metal and its applications. Chem Educ 1997; 2: 1-21.
        [http://dx.doi.org/10.1007/s000897970155a]

[8]   Buehler WJ, Gilfrich JV, Wiley RC. Effect of low-temperature phase changes on the mechanical properties of alloys near composition TiNi. J Appl Phys 1963; 34: 1475-7.
        [http://dx.doi.org/10.1063/1.1729603]

[9]   Wu MH, Schetky LM. Industrial applications for shape memory alloys. Proceedings of the international conference on shape memory and superelastic technologies 2000; 171-82.

[10]   Zider RB, Krumme JF. Eyeglass frame including shape-memory elements U.S. Patent 4,772,112.1988

[11]   Leo DJ, Weddle C, Naganathan G, Stephen JB. Vehicular applications of smart material systems. 5th Annual International Symposium on Smart Structures and Materials Int Soc Opt Photon 1998; 106-16.

[12]   Otsuka K, Ren X. Physical metallurgy of Ti-Ni-based shape memory alloys. Prog Mater Sci 2005; 50: 511-678.
        [http://dx.doi.org/10.1016/j.pmatsci.2004.10.001]

[13]   Ren X, Otsuka K. Origin of rubber-like behaviour in metal alloys. Nature 1997; 389: 579-82.
        [http://dx.doi.org/10.1038/39277]

[14]   Pelton BL, Slater T, Pelton AR. Effects of hydrogen in TiNi. Proceedings of the Second International Conference on Shape Memory and Superelastic Technologies (SMST) 1997.

[15]   Eggeler G, Khalil-Allafi J, Gollerthan S, Somsen C, Schmahl W, Sheptyakov D. On the effect of aging on martensitic transformations in Ni-rich NiTi shape memory alloys. Smart Mater Struct 2005; 14: 186.
[http://dx.doi.org/10.1088/0964-1726/14/5/002]

[16]   Bram M, Ahmad-Khanlou A, Heckmann A, Fuchs B, Buchkremer HP, Stöver D. Powder metallurgical fabrication processes for NiTi shape memory alloy parts. Mater Sci Eng A 2002; 337: 254-63.
[http://dx.doi.org/10.1016/S0921-5093(02)00028-X]

[17]   Kato M, Pak HR. Thermodynamics of Stress-Induced First-Order Phase Transformations in Solids. Phys Status Solidi, B Basic Res 1984; 123: 415-24.
[http://dx.doi.org/10.1002/pssb.2221230203]

[18]   Flyvholm MA, Nielsen GD, Andersen A. Nickel content of food and estimation of dietary intake. Z Lebensm Unters Forsch 1984; 179(6): 427-31.
[http://dx.doi.org/10.1007/BF01043419] [PMID: 6395539]

[19]   Haider W, Munroe N. Assessment of corrosion resistance and metal ion leaching of nitinol alloys. J Mater Eng Perform 2011; 20(4): 812-5.
[http://dx.doi.org/10.1007/s11665-011-9892-5] [PMID: 21666858]

[20]   Briceño J, Romeu A, Espinar E, Llamas JM, Gil FJ. Influence of the microstructure on electrochemical corrosion and nickel release in NiTi orthodontic archwires. Mater Sci Eng C 2013; 33(8): 4989-93.
[http://dx.doi.org/10.1016/j.msec.2013.08.024] [PMID: 24094215]

[21]   Danilov A, Kapanen A, Kujala S, *et al.* Biocompatibility of austenite and martensite phases in NiTi-based alloys. J Phys IV 2003; 112: 1117-20.
[http://dx.doi.org/10.1051/jp4:20031078]

[22]   Habijan T, Glogowski T, Kühn S, *et al.* Can human mesenchymal stem cells survive on a NiTi implant material subjected to cyclic loading? Acta Biomater 2011; 7(6): 2733-9.
[http://dx.doi.org/10.1016/j.actbio.2011.02.022] [PMID: 21345390]

[23]   Peitsch T, Klocke A, Kahl-Nieke B, Prymak O, Epple M. The release of nickel from orthodontic NiTi wires is increased by dynamic mechanical loading but not constrained by surface nitridation. J Biomed Mater Res A 2007; 82(3): 731-9.
[http://dx.doi.org/10.1002/jbm.a.31097] [PMID: 17326228]

[24]   Poon RW, Ho JP, Liu X, *et al.* Improvements of anti-corrosion and mechanical properties of NiTi orthopedic materials by acetylene, nitrogen and oxygen plasma immersion ion implantation. Nucl Instrum Methods Phys Res Sect B. Beam Interactions with Materials and Atoms 2005; 237: 411-6.

[25]   Thierry B, Tabrizian M, Trepanier C, Savadogo O, Yahia L. Effect of surface treatment and sterilization processes on the corrosion behavior of NiTi shape memory alloy. J Biomed Mater Res 2000; 51(4): 685-93.
[http://dx.doi.org/10.1002/1097-4636(20000915)51:4<685::AID-JBM17>3.0.CO;2-S]
[PMID:10880117]

[26]   Bansiddhi A, Sargeant TD, Stupp SI, Dunand DC. Porous NiTi for bone implants: a review. Acta Biomater 2008; 4(4): 773-82.
[http://dx.doi.org/10.1016/j.actbio.2008.02.009] [PMID: 18348912]

[27]   Bogdanski D, Epple M, Esenwein SA, *et al.* Biocompatibility of calcium phosphate-coated and of geometrically structured nickel-titanium (NiTi) by *in vitro* testing methods. Mater Sci Eng A 2004; 378: 527-31.
[http://dx.doi.org/10.1016/j.msea.2003.11.071]

[28]   Liu X, Wu S, Yeung KW, *et al.* Relationship between osseointegration and superelastic biomechanics

in porous NiTi scaffolds. Biomaterials 2011; 32(2): 330-8.
[http://dx.doi.org/10.1016/j.biomaterials.2010.08.102] [PMID: 20869110]

[29]   Rhalmi S, Charette S, Assad M, Coillard C, Rivard CH. The spinal cord dura mater reaction to nitinol and titanium alloy particles: a 1-year study in rabbits. Eur Spine J 2007; 16(7): 1063-72.
[http://dx.doi.org/10.1007/s00586-007-0329-7] [PMID: 17334794]

[30]   Gu YW, Tay BY, Lim CS, Yong MS. Characterization of bioactive surface oxidation layer on NiTi alloy. Appl Surf Sci 2005; 252: 2038-49.
[http://dx.doi.org/10.1016/j.apsusc.2005.03.207]

[31]   Wu S, Liu X, Chan YL, *et al.* Nickel release behavior, cytocompatibility, and superelasticity of oxidized porous single-phase NiTi. J Biomed Mater Res A 2007; 81(4): 948-55.
[http://dx.doi.org/10.1002/jbm.a.31115] [PMID: 17252548]

[32]   Wang XX, Hayakawa S, Tsuru K, Osaka A. Improvement of bioactivity of H(2)O(2)/TaCl(5)-treated titanium after subsequent heat treatments. J Biomed Mater Res 2000; 52(1): 171-6.
[http://dx.doi.org/10.1002/1097-4636(200010)52:1<171::AID-JBM22>3.0.CO;2-O]
[PMID: 10906689]

[33]   Matsuno H, Yokoyama A, Watari F, Uo M, Kawasaki T. Biocompatibility and osteogenesis of refractory metal implants, titanium, hafnium, niobium, tantalum and rhenium. Biomaterials 2001; 22(11): 1253-62.
[http://dx.doi.org/10.1016/S0142-9612(00)00275-1] [PMID: 11336297]

[34]   Ng KW, Man HC, Yue TM. Characterization and corrosion study of NiTi laser surface alloyed with Nb or Co. Appl Surf Sci 2011; 257: 3269-74.
[http://dx.doi.org/10.1016/j.apsusc.2010.10.154]

[35]   He XM, Rong LJ, Yan DS, Li YY. TiNiNb wide hysteresis shape memory alloy with low niobium content. Mater Sci Eng A 2004; 371: 193-7.
[http://dx.doi.org/10.1016/j.msea.2003.11.044]

[36]   Andreasen GF, Hilleman TB. An evaluation of 55 cobalt substituted Nitinol wire for use in orthodontics. J Am Dent Assoc 1971; 82(6): 1373-5.
[http://dx.doi.org/10.14219/jada.archive.1971.0209] [PMID: 5280052]

[37]   Torrisi L. The NiTi superelastic alloy application to the dentistry field. Biomed Mater Eng 1999; 9(1): 39-47.
[PMID: 10436852]

[38]   Idelsohn S, Peña J, Lacroix D, Planell JA, Gil FJ, Arcas A. Continuous mandibular distraction osteogenesis using superelastic shape memory alloy (SMA). J Mater Sci Mater Med 2004; 15(4): 541-6.
[http://dx.doi.org/10.1023/B:JMSM.0000021135.72288.8f] [PMID: 15332632]

[39]   Torrisi L, Di Marco G. Physical characterisation of endodontic instruments in NiTi alloy. Mater Sci Forum Trans Tech Pub 2000; 327: 75-8.
[http://dx.doi.org/10.4028/www.scientific.net/MSF.327-328.75]

[40]   Parashos P, Messer HH. The diffusion of innovation in dentistry: a review using rotary nickel-titanium technology as an example. Oral Surg Oral Med Oral Pathol Oral Radiol Endod 2006; 101(3): 395-401.
[http://dx.doi.org/10.1016/j.tripleo.2005.02.064] [PMID: 16504875]

[41]   Grummon DS, Shaw JA, Foltz J. Fabrication of cellular shape memory alloy materials by reactive eutectic brazing using niobium. Mater Sci Eng A 2006; 438: 1113-8.
[http://dx.doi.org/10.1016/j.msea.2006.03.113]

[42]   Grummon DS, Shaw JA, Gremillet A. Low-density open-cell foams in the NiTi system. J Appl Phys Lett 2003; 82: 2727-9.
[http://dx.doi.org/10.1063/1.1569036]

[43]   Prima SB, Tret'Yachenko LA, Petyukh VM. Phase relations in the Ti-TiNi-NbNi-Nb region of the

ternary system Ti-Nb-Ni. Powder Metall Met Ceramics 1996; 34: 155-60.
[http://dx.doi.org/10.1007/BF00559560]

[44]   Grummon D, Low KB, Foltz J, Shaw J. A new method for brazing nitinol based on the quasibinary TiNi-Nb system. 48th AIAA/ASME/ASCE/AHS/ASC Structures, Structural Dynamics, and Materials Conference American Institute of Aeronautics and Astonautics 2007; 1741.

[45]   Grummon D, Low KB, Foltz J, Shaw J. A New Method for Brazing Nitinol Based on the Quasibinary TiNi-Nb System Aiaa/asme/asce/ahs/asc Structures. Structural Dynamics, and Materials Conference 2015.

[46]   Wang L, Wang C, Lu W, Zhang D. Superelasticity of NiTi-Nb metallurgical bonding *via* nanoindentation observation. Mater Lett 2015; 161: 255-8.
[http://dx.doi.org/10.1016/j.matlet.2015.08.089]

[47]   Wang L, Wang C, Zhang LC, Chen L, Lu W, Zhang D. Phase transformation and deformation behavior of NiTi-Nb eutectic joined NiTi wires. Sci Rep 2016; 6: 23905.
[http://dx.doi.org/10.1038/srep23905] [PMID: 27049025]

[48]   Wang L, Wang C, Dunand DC. Microstructure and strength of NiTi-Nb eutectic braze joining NiTi wires. Metall Mater Trans, A Phys Metall Mater Sci 2015; 46: 1433-6.
[http://dx.doi.org/10.1007/s11661-015-2781-z]

[49]   Neurohr AJ, Dunand DC. Shape-memory NiTi with two-dimensional networks of micro-channels. Acta Biomater 2011; 7(4): 1862-72.
[http://dx.doi.org/10.1016/j.actbio.2010.11.038] [PMID: 21130189]

[50]   Bil C, Massey K, Abdullah EJ. Wing morphing control with shape memory alloy actuators. J Intell Mater Syst Struct 2013; 24: 879-98.
[http://dx.doi.org/10.1177/1045389X12471866]

[51]   Hartl DJ, Lagoudas DC. Aerospace applications of shape memory alloys. Proc Inst Mech Eng Part G J Aerosp Eng 2007; 221: 535-52.
[http://dx.doi.org/10.1243/09544100JAERO211]

[52]   Morgan NB. Medical shape memory alloy applications-the market and its products. Mater Sci Eng A 2004; 378: 16-23.
[http://dx.doi.org/10.1016/j.msea.2003.10.326]

[53]   Noonai N, Khantachawana A, Kaewtathip P, Kajornchaiyakul J. Improvement of Mechanical Properties and Transformation Behavior of NiTi Drawn Wires for Orthodontics Applications. Adv Mat Res 2012; 378: 623-7.

[54]   Bansiddhi A, Sargeant TD, Stupp SI, Dunand DC. Porous NiTi for bone implants: a review. Acta Biomater 2008; 4(4): 773-82.
[http://dx.doi.org/10.1016/j.actbio.2008.02.009] [PMID: 18348912]

[55]   Biswas A. Porous NiTi by thermal explosion mode of SHS: processing, mechanism and generation of single phase microstructure. Acta Mater 2005; 53: 1415-25.
[http://dx.doi.org/10.1016/j.actamat.2004.11.036]

[56]   Shu XY, Lu SQ, Li GF, Liu JW, Peng P. Nb solution influencing on phase transformation temperature of Ni 47 Ti 44 Nb 9, alloy. J Alloys Compd 2014; 609: 156-61.
[http://dx.doi.org/10.1016/j.jallcom.2014.04.165]

[57]   Uchida K, Shigenaka N, Sakuma T, Sutou Y, Yamauchi K. Effect of Nb Content on Martensitic Transformation Temperatures and Mechanical Properties of Ti-Ni-Nb Shape Memory Alloys for Pipe Joint Applications. Mater Trans 2007; 48: 445-50.
[http://dx.doi.org/10.2320/matertrans.48.445]

[58]   Shi H, Frenzel J, Martinez GT, *et al.* Site occupation of Nb atoms in ternary Ni-Ti-Nb shape memory alloys. Acta Mater 2014; 74: 85-95.
[http://dx.doi.org/10.1016/j.actamat.2014.03.062]

[59]   Piao M, Miyazaki S, Otsuka K, Nishida N. Effects of Nb addition on the microstructure of Ti-Ni alloys. Mater Trans, JIM 1992; 33: 337-45.
[http://dx.doi.org/10.2320/matertrans1989.33.337]

[60]   Jani JM, Leary M, Subic A. Mark AGA review of shape memory alloy research, applications and opportunities. Mater Des 2014; 56: 1078-113.
[http://dx.doi.org/10.1016/j.matdes.2013.11.084]

[61]   Bil C, Massey K, Abdullah EJ. Wing morphing control with shape memory alloy actuators. J Intell Mater Syst Struct 2013; 24: 879-98.
[http://dx.doi.org/10.1177/1045389X12471866]

CHAPTER 7

Surface Modification of Biomedical Titanium Alloys

Zihao Ding[1], Liqiang Wang[1,*], Chengjian Zhang[1] and Lai-Chang Zhang[2]

[1] *State Key Laboratory of Metal Matrix Composites, Shanghai Jiao Tong University, No. 800 Dongchuan Road, Shanghai200240, PR China*

[2] *School of Engineering, Edith Cowan University, 270 Joondalup Drive, Joondalup, Perth, WA, 6027, Australia*

Abstract: Possessing excellent bio-compatibility and mechanical performance, Ti-6Al-4V (TC4) alloy is widely used as implant. Aiming at solving the problem of poor surface wear properties and improving bio-compatibility of TC4 alloy, friction stir processing(FSP) is applied to it filled with TiO_2 powder in the groove to realize surface modification and build nano-sized composite biomedical material. Change in microstructure and its relationship with mechanical performance such as hardness will be discussed. A series of experiments in biology, including cytotoxicity test, cell culture, adhesion, proliferation and alkaline phosphatase activity is carried out to verify bio-compatibility of the material, compared with original TC4. The improved material is expected to provide a better environment for cells to grow.

Keywords: Bio-compatibility, Biomedical titanium alloy, Friction stir process, Nano-sized composite layer, Surface modification.

INTRODUCTION

Because of the excellent properties of high strength, sufficient stiffness and strength-to-weight ratio, titanium alloys, have widespread been used in aerospace, chemical, nuclear industries and especially, biomedical application as "load-bearing" implants connected to bony tissue [1, 2]. The medical products, made by titanium alloy, such as artificial joints, stents, dental implantations, are wildly used in clinical diagnosis, treatment, repair, implantation or enhancing the body tissues and organs' function, which bring an irreplaceable effect in therapy [3]. Compared with other materials which are used in biomedical field earlier, such as pure titanium stainless steel, titanium alloys earn their advantage because of

* **Corresponding author Liqiang Wang:** State Key Laboratory of Metal Matrix Composites, School of Materials Science and Engineering, Shanghai Jiao Tong University, No. 800 Dongchuan Road, Shanghai 200240, P.R. China; Tel: 8602134202641; Fax: 8602134202749; E-mail: wang_liqiang@sjtu.edu.cn

excellent bio-compatibility and corrosion resistance make titanium [3]. Artificial joint replacement is the most effective method of treatment for end-stage bone joint disease, which can observably recover and improve the motor ability of the patients. With the coming of the aging society and the improvement of people's living standard, the demand for the medical implants will continue to grow.

Among all of the titanium alloys, Ti-6Al-4V (TC4) alloy is the most frequently and successfully used α + β titanium alloy in biomaterial field due to its many favorable properties. Since the very first attempt in the early 1950s, TC4 titanium alloy has become a backbone material for several fields in a relatively short time, especially in biomedicine in place of pure titanium [4].

However, titanium alloys present certain disadvantages as biomedical material that needed to be solved. Ti-6Al-4V alloy presents poor surface wear properties in long term service in human body which limits the performance and service life [5]. Besides, according to former research, tiny titanium particles are found in soft tissues adjacent to implants accompanied by inflammation which are caused by both abrasion and corrosion in human body that make them adrift from the surface of implants [6, 7]. Actually, in study of mechanism of electrochemical reaction of titanium alloys, when exposed to solution of 3.5% sodium chloride, titanium element in the alloy tends to be converted into TiO_2 and TiO_{1+x} (which means non-stoichiometric oxide) that is relatively more stable but it also changes the properties of the surface of alloy [8]. Therefore, appropriate measures should be taken to improve wear property, corrosion resistance and bio-compatibility of titanium alloys to make them better meet requirements of implant.

Among all methods of improvement, surface modification is the most convenient way to realize all the objectives mentioned above, since it only changes the structure and properties of the surface part of the material. In terms of biocompatibility it helps to make the fixation between prosthesis and the bone rely more on bone tissue, and improve the bone integration on the bone tissue implant interface and the long-term stability of biological prosthesis. And this process can also promote the early osseointegration between the implant bone's interface and the bone, which has undoubted significance to prevent prosthesis loosening.

Friction Stir Processing (FSP) is attempted to improve surface performance of titanium alloys in this study. FSP was developed by Mishra *et al.* for microstructural modification based on the basic principles of friction stir welding (FSW) [9, 10]. FSW is a new solid-state joining technique invented at The Welding Institute (TWI) (Cambridge, United Kingdom) in 1991 which is considered as the most significant development in metal joining in a decade, and has been successfully utilized to produce joints in titanium and its alloys [11, 12]. Normal castings have poor comprehensive mechanical properties because of the flaws such as porosity, shrinkage cavity and dendrite after the welding. These flaws are very hard to eliminate totally by traditional heat treatment or chemical modification methods. However, friction stir processing avoids such problems through its solid-state soldering. The most important part of a friction stir processing system is its rotating tool with pin and shoulder. Friction stir processing is realized by the high-speed rotation with the pin inside the material and movement of the rotating tool. During the processing, the rotating tool rotates at high speed and then slowly comes into the work piece, until the shaft shoulder and the surface of the work piece contact closely. Then, the threaded pin comes into the inner material to rub and stir. The shear heating caused by it makes the metal around the rotating tool soft, and causes plastic flowing. As the rotating tool goes in the processing direction at a certain speed, the stirring and friction of the threaded pin make the heat input the friction-stir-processed area continually, which cause the severe plastification on material's surface, so that the grains emerge severe deformation and dynamic recrystallization to realize the improvement of the surface properties of materials [13].

Friction stir processing can modify the microstructure of materials, including refining the matrix grain, eliminating defectives of casting holes and breaking dendrite and the second phase without changing the overall component's shape and size [14]. All these changes lead to the increase of microhardness, strength, corrosion resistance and fatigue performance of materials, and finally improve its surface properties. The economic and technological benefits of FSP have been well recognized in various engineering materials. Because of the high quality, efficiency, low energy consumption and no pollution of friction stir processing, it is widely used in aerospace, automotive and other fields, which has become a hot spot of research at new environment-friendly processing technology in the whole world. Generally speaking, the FSP technique is emerging as a very effective solid-state processing technique that can provide localized modification and control of microstructures in the near-surface layers of processed metallic components [15].

Friction stir processing (FSP) is a type of surface modification technology with many advantages:

1. From the view of processing results, first of all, the shear friction heat, produced by the shaft shoulder is the main heat input in the processing. Secondly, although the processing area is small, there is a severe plastic deformation. Machining area of the material has the plastic flow due to the friction heat. In the deformation process, the large grains are broken into

obvious fine grains and ultrafine grains. At the same time, the deformation zone is so small that the work piece has good dimensional stability and repeatability [3].

2. As for machining efficiency, this technology is simpler, less time consuming, and has high production efficiency. First of all, as FSP is a rapid processing, which can not only be used in single channel, but a double one, so that we can process the double sides of the metal plate in the same time, with using less energy, and spending less time. Secondly, the pretreatment of friction stir processing does not require complicated heat treatment process. In addition, if there are no special requirements, it does not need the protection gas, surface cleaning, and other solvent requirements.

3. FSP has low cost. Although the initial investment cost is high, pretreatment is simple and costs little. Especially in the continuous processing, the save of the cost is more obvious.

4. From the perspective of environmental protection, FSP saves materials, and has low energy consumption with high efficiency. The process of it is safe without producing pollution, dust *etc*.

FSP has been applied to realize surface modification, refine microstructure, and prepare metal matrix composite materials. According to the research, the technics parameters, including the shape of stir head, the speed of processing and rotation, all have significant effects on the morphology of the surface modification. Recently, because of the tremendous progress in FSW/FSP, more and more people are studying about the FSP on high melting point materials such as titanium, Ni-Ti alloys *etc*. Han [16] studied on the the porosity and morphology of A1050 porous aluminum, which is processed by FSP in different speed of the rotating tool. The results show that, using the minimum times of processing can improve the productivity. Feng [17] studied on Mg-Al-Zn processed by FSP and aging casting. FSP made the grains into obvious fragmentation. Reticular eutectic phase distributes on the grain boundaries. FSP could also eliminate trachoma, porosity and other casting defects. And the castings obtained denser microstructure, higher hardness and better tensile properties after processing. The conclusion is that FSP combined with aging casting can improve the mechanical properties of Mg - Al - Zn, in a simple and effective method. Kitamura [18] studied the microstructure and mechanical properties of 2 mm thick TC4 (Ti-6Al-4V) titanium alloy plant, which was processed under FSW in different parameters and thermal cycling. When the welding temperature is higher than the β-phase transition temperature, the size of pioneer β-phase grain in stirring zone changes according to the peak temperature, and the size of the layered structure depends on the cooling rate. When the welding temperature is lower than the β-phase transition temperature, the microstructure in stirring area is the primary α-phase equal axis crystal microstructure with higher tensile strength than the matrix. Li

[19] studied on the surface modification of Ti-6Al-4V alloy, processed under the different parameters *via* FSP. The friction stir processing layer without any defects is about 2.5 mm and the final β-phase zone is composed of acicular α-phase, grain boundary α-phase and martensitic α'-phase. According to the change on the microstructure of Ti-6Al-4V, they clarified the relationship between the processing parameters and the microstructure's characteristics. They discovered that the low peak temperature with fast cooling speed will lead to the formation of fine acicular β-phase region, α-phase and acicular martensite α'-phase. Compared with the mechanical properties of matrix, the hardness and the wear performance of FSPed surface layer are largely improved. Ma [20] has studied the microstructure and properties of TA2 pure titanium, which is processed by FSP. The results showed that after processing, the grains in the stir zone were greatly refined by severe plastic deformation. After processing, the anti-friction and wear performance of industrial pure titanium was obviously improved, too.

Besides improvement of microstructure and mechanical properties, FSP also provides probability to add second-phase particle into raw material because of its huge amount of heat input and severe plastic deformation. Titanium oxide, TiO_2, helps retain the natural forms of proteins and also minimizes unwanted platelet activation because of passive surface and chemical stability which makes it ideal compatible coatings on artificial bone and dental implants [21, 22]. According to various researches, chemical surface treatments, coatings, or anodization of titanium to form TiO_2 are applied to increase the bioactivity of titanium-based implants [23 - 28]. However, there exist several problems using such measures to generate titanium oxide in order to improve bio-compatibility, including lack of sufficient adhesion between TiO_2 and base material that could cause oxide layer flaking not only during service in human body but also in surgical implantation procedures. To solve the problem, we take friction stir processing into consideration, the mechanism of which is totally different from methods mentioned before. While FSP can improve the surface properties of Ti-6Al-4V, it also makes uniform distribution and dispersion of TiO_2 in Ti-6Al-4V alloy as base material possible in this study.

The aim of this chapter is describe how we fabricate a kind of bioactive titanium alloy comprised of Ti-6Al-4V alloy as base material and nano-sized composite layer of TiO_2 and TC4 as surface with topology structure through FSP technology. Observation of microstructure and experiments of mechanical property, bio-compatibility and corrosion resistance will be done on samples with different filling amount of TiO_2 as well as Ti-6Al-4V alloy without modification, in order to figure out the effects of TiO_2 and FSP technology.

MATERIALS AND METHODS

Materials and FSP System

5 mm thick rolled plates of Ti-6Al-4V alloy (5.5-6.8 wt% Al, 3.5-4.5 wt% V, 0.3 wt% Fe, 0.1 wt% C and bal. Ti) with annealing treatment are used as base material. Wire cutting is used in this study to groove on the surface of plates for filling of TiO_2 powder. To ensure uniform distribution of TiO_2, the width of each groove and separation distance is constant. The depth of groove is the only variable to determine the amount of TiO_2 filled. In this study, three grooves' depths (1mm, 1.5mm and 2mm) are applied. Figs. (**1a** and **b**) show the appearance of plates with grooves and various parameters, including the width and separation distance of slot. TiO_2 powder is filled into the groove until it filled the groove. The TiO_2 powder utilized in the study is white amorphous powder in chemical pure standard. The average diameter of the powder is 200 nm.

The key part of the FSP system is a stir-tool of Tungsten carbide (WC)-based alloy composed of a 15mm-diameter concaved shoulder and a 7mm-diameter unthreaded pin which is 2mm in height and tilted by 2.5° responsible for producing an FS-processed surface layer on the substrate. During FSP, stir-tool moves forward at a constant velocity and keeps parallel to the upper-surface of the plate with unthreaded pin rotating which will cause fierce plastic deformation and combination of Ti-6Al-4V alloy and TiO_2. Meanwhile, shielding gas is provided to prevent oxidation of alloy under high temperature. Fig. (**1c** and **d**) is a view of FSP system and its operational principles. Among parameters of FSP, the forward velocity of stir-tool is 50 mm/min and two different speed rates of rotation of unthreaded pin are utilized which are 225r/min and 375r/min respectively. All samples experience single-pass FSP.

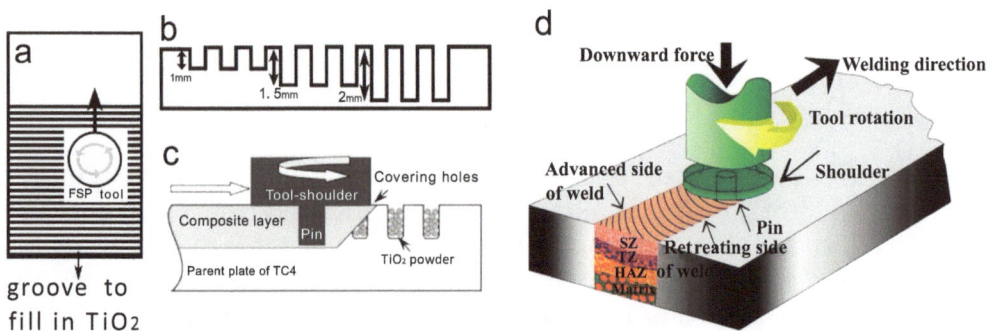

Fig. (1). Materials and FSP system: (**a**) vertical view of TC4 plate and the groove; (**b**) grooves with different depth; (**c**) composite layer and parent plate of TC4 and (**d**) schematic diagram of FSP procedure.

Microstructure Observation and Composition Analysis

Before the metallographic analysis, all the samples are cut, polished and etched. When polishing, we use 600#, 1000# and 1500# sandpapers successively. Then polish samples until the surface has no scratches. The corrosion solution of in the study is composed of 1:2:50 HF, HNO_3, H_2O. The time to wipe the surface with corrosive liquid is about 30 seconds. After corrosion, we can obviously observe the stirring transition zone and the crescent. Firstly, observe all the samples under the metallographic microscope, and take metalloscope photos. Then compare the metallographs with different processing parameters, and do the microstructure analysis. High-magnification microstructures were analyzed using a Quanta 200 scanning electron microscope (SEM) (FEI Company, The Netherlands) with an energy dispersive X-ray spectroscope (EDS) (EDAX Inc., USA) to determine the chemical compositions. To compare difference of composition between samples and unprocessed Ti-6Al-4V alloy we also use XRD technology as a supplementary means. Using the transmission electron microscope, we can observe and analyze microstructure and defects of stir zone. Use wire cutting to cut off stir zone into a 500µm slice, affix it to the rubber, grind it to about 100µm with extremely fine sandpaper, then grind it to 80µm by hand, then do the double electrolytic polishing. Firstly put the 3mm diameter wafers made by punching and shearing method into the sample clip, then use electrolyte to polish the sample through jetting the center of the wafer. The electrolyte formula was: 3% perchloric acid +97% alcohol, set temperature in -25 degrees, voltage in 75V. Then observe and analyze the dispersion and morphology of the nanoparticles under JEM 2100, measure and assess the size of the nanoparticles, and observe the microstructure and defects.

Nanoindentation

Nanoindentation applies a controlled load to the surface of material which can induce local surface deformation in order to measure some mechanical properties. During the process of loading and unloading, load and displacement are monitored continuously and recorded in a text to plot a certain curve. Hardness and elasticity modulus can be calculated according to the unloading curves using equations based on elastic contact theory. Samples of three parameters (225-2, 300-2, 375-2) were mounted with a exposure of lateral surface. Before test, samples were disposed by standard process and vibration polishing aiming at eliminating the residual stress on the surface. Indentations were created in a vertical line from stir area to substrate area with a 0.5mm and 4 points in each area using the HVS-1000 Testing System (Nanjing HVS Inc., China) at 0.2 kg load and 15s indentation time. Hardness and modulus were calculated then to obtain a comparison among different areas and among different parameters.

In Vitro Biocompatibility Test

The test of biocompatibility is composed of a series of experiment conducted *in vitro* to characterize the ability of TC4-TiO$_2$ composite material to promote cell proliferation and differentiation:

(1) Cytotoxicity Test

The specimens were immersed into minimum essential medium(MEM) for 72h under a cell condition of 37 °C and 5% CO$_2$ for preparing extracts. Then the extracts were diluted into different concentration of 25%, 50%, 75% and 100% respectively for cytotoxicity test. Murine fibroblast cells (L-929) obtained from Cell Bank, Chinese Academy of Sciences, were used as the cell line to identify cytotoxicity. L929 were seeded on 96-well plates with a cell density of 10^4 cells as well as 100μl MEM medium per well. After 24 h incubation under the condition of 37 °C and 5% CO$_2$, the medium was removed and 100μl leaching solution of the test with the extracts prepared before(25%,50%,75%,100%), empty group, negative as well as positive control were added with another 24 h incubation. After this the medium was removed again and 50 MTT solution were poured into each well. The MTT solution with replaced with isopropanol (100μl/well) after 3 h incubation. At last, the absorbance of samples were measured at 570nm using microplate reader after the vibration of plates.

(2) Cell Culture

In the following tests, MG-63 human osteosarcoma cells obtained from Cell Bank, Chinese Academy of Sciences, were chosen as the cell line. For each experiment, MG-63 were seeded directly on titanium alloy surfaces in 48-well plate with a density of 2000 cells/500μl medium/well in adding high glucose DMEM. Then they were incubated for the time specifically each test requires.

(3) Cell Adhesion

The cultivation was terminated after the samples were incubated for 12 h at 37 °C in an atmosphere of 5% CO$_2$. After cleaning three times in phosphate buffer solution(PBS), divided the samples in two groups. In one group, each well was dyed using DIPI and placed under fluorescence microscope to measure the number of cells so as to gain a conclusion about the adhesion ability. In another group samples were dried for 4 h and then were sprayed. Finally, the samples were put into SEM to observe the cell adhesion form and gain a comparison.

(4) Cell Proliferation

The samples were evaluated after1, 4, 8 and 16 d cell culture under the condition of 37°C and 5% CO_2. After the samples were washed with phosphate buffer solution (PBS) twice, CCK8 was added with 100µl/well with another 4 h incubation. Finally, from each well 50µl solution were extracted to another 96-well plate and the optical densities of samples were measured at 570nm under microplate reader.

(5) Alkaline Phosphatase Activity

The cultivation was ended after 1, 4, 8 and 16 d culturing and 0.05%Triton X-100 was added into each well in order to lyse osteoblast and prepare cell lysis buffer (4°C, 1h). From each sample, 50µl solution was extracted and incubated at 37°C for 0.5h with an addition of 100µl DEA and 50µl PNPP. Eventually, the reaction was terminated with the help of 100µl NaOH (1mol/L) and the optical densities of samples were measured at 405nm by 318MC ELIASA according to which the osteoblast alkaline phosphatase levels can be calculated. In order to obtain the total amount of protein, cells were cracked by three freeze-thaw cycles after 4, 8 and 16 d cell culturing and three times cleaning in PBS. 1ml distilled water was added into each well and from each well three copies of 100µl solution were transferred into 96-well plate in different well as the standard group. 100µl protein marker and 200µl Protein determination reagent were added into each standard well with 0.5 h incubation at 37°C. In addition, 100µl normal saline was added into empty well. The empty well was set to zero and optical densities of samples were measured at 570nm to get to the total concentration of protein. Eventually, the ratio of alkaline phosphatase level and total concentration of protein is alkaline phosphatase activity.

(6) Mineralization Assay

The Alizarin Red assay was used to determine the mineralization ability of osteoblasts [29]. For this purpose, the cells were fixed using 4% paraformaldehyde after 8 and 16 d of cell culturing and twice cleaning in PBS. The cells were dyed by alizarin red (40mM) for 20mins at room temperature and double distilled water was used twice to wash the cells both before and after dye. At last 10% cetane pyrazole chloride was used to terminate the reaction and the solution was transferred into another 96-well plate in order to measure the optical densities at 590 nm.

Corrosion Resistance

In practical situation corrosion gradually occurs to titanium alloy, leading to decline of mechanical properties. In another word, the better resistance to corrosion titanium alloy has, the longer its service life will be. Through polarization curve and AC impedance test using an electrochemical workstation potential-time chart and Tafel plot can be achieved, according to which we can figure out resistance to corrosion of samples and have a better understanding of the whole corrosion process.

To simulate internal environment following solutions are used in polarization curve and AC impedance test: Hank's solution and PBS (phosphate buffered solution), which are prepared according to standard process.

RESULTS AND DISCUSSIONS

Microstructure

Fig. (**2**) shows the microstructure in different zones of the sample. In Fig. (**2a**), the morphology can be divided into three parts from the longitudinal section after friction stir processing, which are the stirring zone (SZ), the transition zone (TZ) and the base material (BM). The thickness of SZ is about 1.5mm. It should be noticed that two sides are asymmetric because during friction stir processing softened material ahead the traveling tool shoulder migrates from the front at the advancing side (AS, where the tool rotating direction is the same with the traveling direction) to the rear at the retreating side (RS, where the tool rotating direction is opposite against the traveling direction) [30 - 32]. The difference is that at AS the space next to shoulder can be filled quickly by flow of metal due to same direction of forwarding and rotating. According to Fig. (**2a**), inclination angle of AS (46.7°) is larger than that of RS (36.6°). Fig. (**2b**) shows the microstructure at low magnification. Because the rotating pin is the direct input heat source, temperature gradient exists from SZ to BM. Thus, the size of grains decreases gradually from BM to SZ because of the plastic deformation induced by the tool rotation and traveling behaviors. Since the surface is under a better heat dissipation condition and is most affected by heat input, SZ shows the most effective recrystallization after friction stir processing. Moreover, a great deal of plastic deformation and heat input contribute to the uniform distribution of TiO_2 particles and to refining TiO_2 particles into nanoscale.

At higher magnification, the structure of material and the distribution of TiO_2 can be observed clearly. Fig. (**2c-e**) show SZ, TZ and BM of Ti-6Al-4V. BM is composed of α phase and relatively small amounts of grain-boundary β phase. According to Ref [32], as the peak temperature during friction stir processing is higher than the α-β phase transformation temperature, a part of α phase transformed into β phase. Thus, β phase in TZ presents as larger grains compared to that in BM. In SZ, a large amount of TiO_2 particles exist in the form of clusters while they are granular in TZ with a much smaller size. According to the observation, the size of TiO_2 particle in TZ is about 200 nm while only clusters can be seen in SZ. Fig. (**2c-e**) also indicate that the content of TiO_2 from SZ to

BM has realized graded distribution.

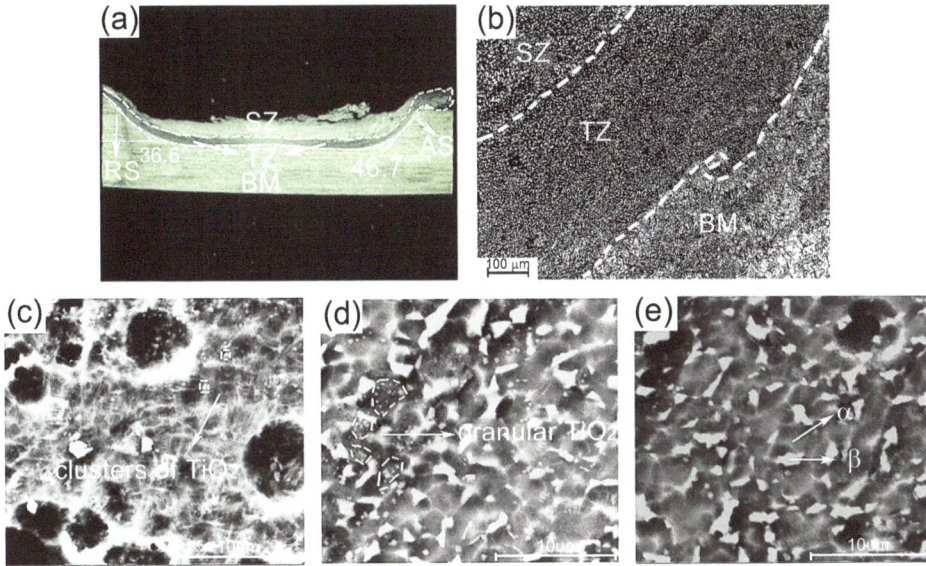

Fig. (2). Microstructure of different zones of the Ti-6Al-4V sample after friction stir processing: (**a**) macroscopic morphology; (**b**) overall microstructure of three zones; (**c**) stirring zone; (**d**) transition zone and (**e**) base material [31].

In order to further elucidate the grain refinement mechanism of the alloy, the microstructure of three zones was further investigated by TEM as shown in Fig. (**3**). Fig. (**3b-c**) show the TiO$_2$ particles with an average size of 10 nm in SZ. The distribution of particles is uniform after plastic deformation. Because of the highest cooling rate in SZ, part of material has formed amorphous state after FSP. In TZ, the volume fraction of amorphous gradually reduces and the material becomes crystallized structure that resembles to the one in BM. Therefore, an obvious interface can be observed in TZ, as shown in Fig. (**3d-e**). From diffraction patter, it is clear that amorphous and crystallized structure coexist. Under the interface the material presents clear lattice structure with atomic layers distances of 0.558 nm while above the interface is annular amorphous structure. It can be seen in the figure that the interface is not smooth but intricate. The division between amorphous and regular crystal lattice reveals the decrease in heat transfer from SZ to TZ, which is not sufficient to cause recrystallization outside the interface. Dislocation wall is formed around the interface, as indicated by the white marks in Fig. (**3e**). Because the effects of plastic flow and deformation in TZ decrease and TiO$_2$ powder is only filled on the surface of the plate of Ti-6Al-4V, the content of TiO$_2$ gradually declines with the increase in depth from sample surface. Fig. (**3f**) shows the microstructure between TZ and BM. It contains elongated α grains and retained β phases which are sporadically distributed between two grains or at junctions.

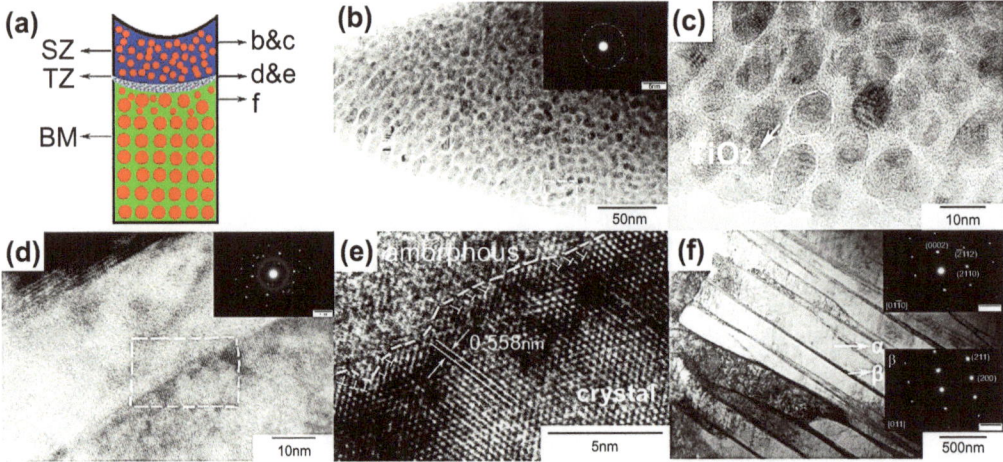

Fig. (3). TEM images of different zones of the Ti-6Al-4V sample after friction stir processing: (**a**) diagrammatic sketch of other figures; (**b**) nanocrystalline in SZ; (**c**) enlarged drawing for b; (**d**) interface between amorphous and crystal; (**e**) enlarged drawing for d and (**f**) structure between TZ and BM [30, 31].

In addition, another effect of friction stir processing is causing the phase transformation and changing the microstructure in matrix. According to phase diagram of Ti-Al-V, the α-β phase transformation temperature ranges during 773-1276K (500-1003°C) and both α and β phases co-exist in this temperature region [33]. Once the temperature in the center of TZ exceeds the transformation temperature of β phase, the material will undergo phase transition and recrystallization [34]. Moreover, because of the slow cooling rate under the atmosphere of protecting gas, SZ and TZ show different microstructures and composites compared with BM. Fig. (**4a**) shows the microstructure of SZ. In SZ, the lamellar α+β phase can be observed in the grains of primary β phase, which is attributed to the high temperature. After friction stir processing, transition from β phase to α phase occurs in condition of slow cooling rate, forming complete lamellar tissues. As is shown in Fig. (**4b**), the dislocation and stacking fault in α phase forms to balance the strain in phase transformation, according to Burgers relationship [35]. Fig. (**4c-d**) shows the microstructure of TZ. Affected by shearing force and stress, the microstructure in TZ also undergoes plastic deformation. Both the lamellar α+β and the equiaxial α phase can be observed in TZ. The equiaxial α phase in TZ originated from two sources. On one hand, TZ doesn't act directly with the stir tool but with SZ mechanically, so that the temperature is lower than that in SZ and a part of primary α phase reserves to the residual primary α phase; on the other hand, the plastic deformation occurs in α+β phase and a part of α grains undergoes dynamic recrystallization to form recrystallized equiaxial α phase [36]. The mechanism of lamellar α+β phase in TZ resembles that in SZ. Therefore, the formation of microstructure in TZ is caused by both the dynamic recrystallization and the phase transformation. Fig. (**4e**) depicts the microstructure of BM. BM reserves more characteristics of structure before friction stir processing as it is away from stir tool compared to TZ and SZ. Fig. (**4f**) shows that in BM there is primary dual phase of α and β, which are mostly equiaxial with respectively bigger size. However, the microstructure change in SZ and TZ causes the stress concentration in the zone, resulting in the dislocation and twinning both inside the grain and on the boundary.

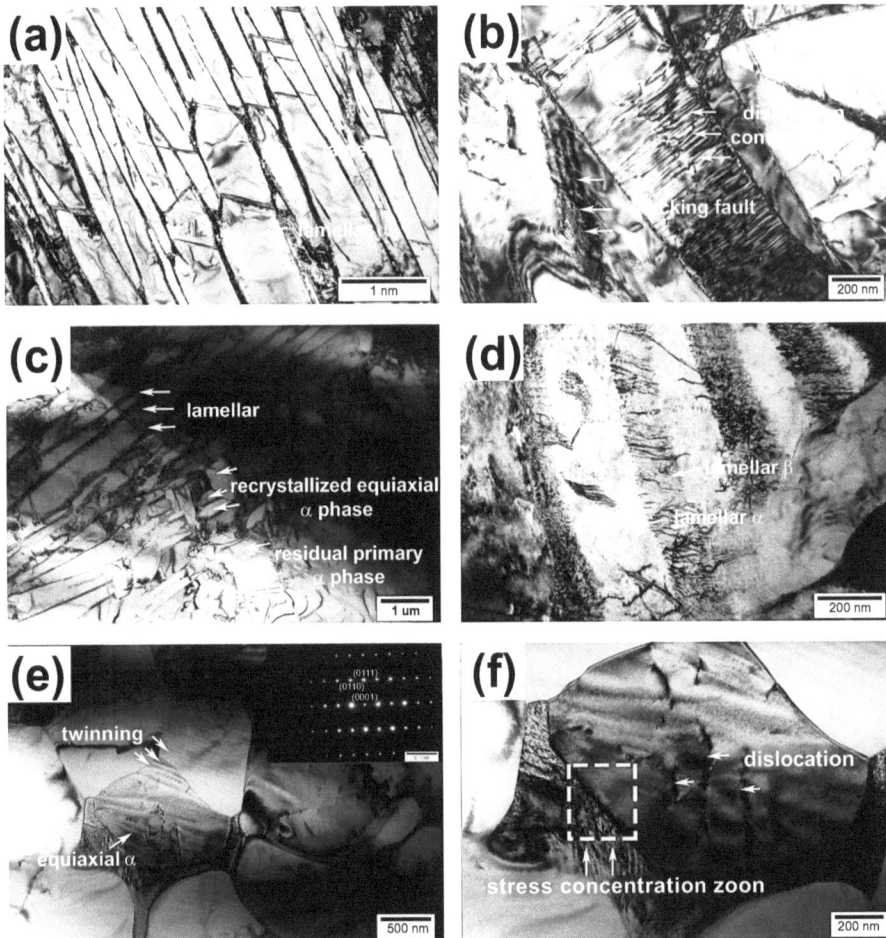

Fig. (4). (a) Lamellar α and β in SZ; (**b**) dislocation and stacking fault in SZ; (**c**) equiaxial and lamellar structure in TZ; (**d**) lamellar a and b in TZ; (**e**) equiaxial structure in BM and (**f**) dislocation and stress concentration zone in BM [31].

The composite layer containing TiO_2 nano-particles was produced on the surface of Ti-6Al-4V by using friction stir processing. Friction stir processing successfully introduces TiO_2 particles into Ti-6Al-4V as second phase, as well as promotes the uniform distribution of TiO_2 particles and the grain refinement in SZ. Friction stir processing also induces the change in microstructure of three zones. In SZ, because the temperature exceeds the transformation temperature of β phase, Ti-6Al-4V undergoes phase transition and recrystallization, forming the lamellar α+β phase. Nanocrystalline TiO_2 particles are distributed in SZ. In TZ, as heat transfer from SZ to TZ decreases, both the lamellar α+β and the equiaxial α phase can be observed. There's also an interface in TZ between amorphous and regular crystal lattice. In BM, microstructure reserves primary equiaxial dual phase of α and β but dislocation and twinning appear because of stress concentration.

Microhardness

Fig. (**5**) shows the results of nanoindentation test. In Fig. (**5a**), the indenters were indicated by 1 to 6 from bottom to top. The insets are the magnified images of indenter 2, 4 and 6 respectively. From the insets, the length of crack caused by indentation increases from 2 to 6, indicating the increase in hardness in some way. Fig. (**5b**) shows the loading-unloading curves of some representative indenters of each sample. There is an overall increasing trend for hardness from BM to SZ as displacement decreased at the same maximum load. Hardness can be calculated according to the improved Oliver-Pharr measurement [37]. The hardness of each test is shown in Fig. (**5c-d**). In Fig. (**5c**), as the sample depth decreases (the row number increases), the hardness of all samples first decreases slightly then increases rapidly, which is evidenced by the variation tendency of hardness in these three zones. This can be attributed to the changes in microstructure. Firstly, in BM, as shown in Fig. (**4f**), the recrystallized equiaxial grains and the pile-up of dislocations can resist the deformation thereby increasing the hardness. Secondly, in TZ, the decreased amount of the defects owing to the increased heat and the increase in α grains size, can lead to a slight decrease in hardness. Thirdly, in lower SZ, nanocrystallines and amorphous TiO_2 are presented as shown in Fig. (**2c**). Based on the well-known Hall-Petch relation, as the size of grain decreases, the strength and hardness increase. Moreover, amorphous particles contributes to the increasing hardness according to previous study [38]. Fourthly, from lower SZ to upper SZ, nanocrystallines become more and much smaller comparing Fig. (**2c**) with Fig. (**2d**). In addition, as shown in Fig. (**2d**), uniformly distributed TiO_2 as a precipitated phase would also increase the hardness. Therefore the hardness of the surface can be greatly enhanced. The hardness of surface can reach 15~19 GPa, much higher than the values of BM (8~10 GPa). The increase in hardness in the 375-1 sample is not as much as that in 375-2. This can be explained by the thin stirring zone in 375-1 samples. However, based on the increasing trend, it can be postulated that the hardness can reach 14 GPa at the surface of 375-1 which is high enough.

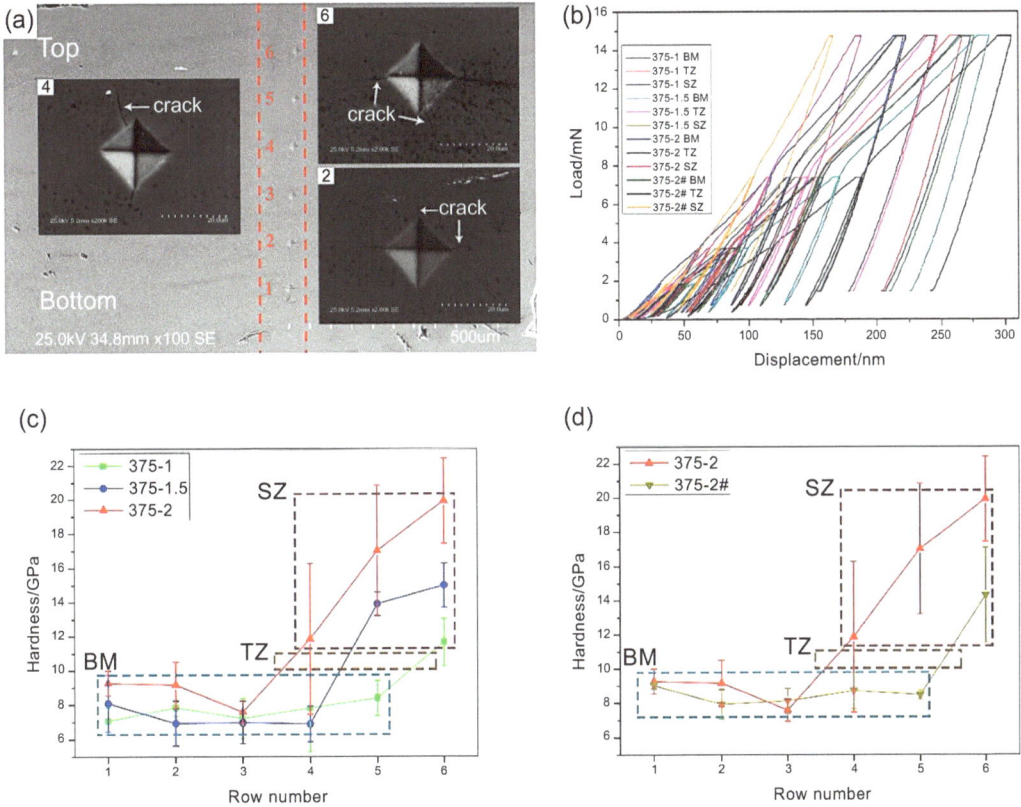

Fig. (5). (**a**) SEM image of the sample 375-1. Numbers 1 to 6 stand for the row number; (**b**) Load-displacement curve of some representative indenters of samples; (**c**) Average hardness of each row in the center area of three samples and (**d**) Average hardness of each row in center and edge area of 375-2 sample. # stands for the edge area. Error bar shows standard errors [30].

From Fig. (**5c**), among the samples with different groove depths, the hardness varies in the same row number (row 4, 5 and 6). This demonstrates that more TiO_2 particles contribute to a higher hardness. In Fig. (**5c**), the hardness of 375-2 sample begins to increase at the area between row 3 and 4 while the hardness of 375-1.5 starts to increase between row 4 and row 5. These row numbers show the locations that the transition zone starts. Therefore, the lower numbers mean larger SZ and TZ. It is seen that as the groove depth increases the stirring zone and transition zone become larger. Fig. (**5d**) compares the hardness between the center area and the edge area in 375-2. Because TZ and SZ have a slope in the edge area as shown in Fig. (**2a**), TZ is in a higher row indenter number. Additionally, SZ is very thin in edge area indicating a lower hardness than that in the center areas. In Fig. (**5d**), TZ of the edge area starts between row 5 and row 6 and the center areas start between row 3 and row 4. The surface hardness in the edge area is also ~5GPa lower than the center areas showing relative weaker FSP effects in edge area than those in center effect. Variation tendency of hardness also verifies the aforementioned discussions that hardness decreases a bit then increase a lot as depths decreases.

The nanosized composites on the surface shows great improvement on microhardness. With the help of nanocrystallines, the surface microhardness of FSPed alloy (15~19GPa) is also larger than those modified by other methods, such as anodic oxidation (6~8GPa) [39]. Concerning that the hardness increases greatly, the wear of the implants should be obviously alleviated. Hopefully, the service life of the implants will also be prolonged.

Biocompatibility Test

(1) Cytotoxicity Assay

As shown in Fig. (**6**), all cell viabilities of the groups are above 100% which indicates there are something benefit for cell in the medium. Nevertheless, TC4 has no cytotoxicity was a recognized fact. So using the TC4's data as the standard. There are no significant difference between TC4 and FSP groups showing that FSP with TiO_2 has no adverse effect on cells. In other words, all of the FSP groups have no cytotoxicity. Generally speaking, in a short time about five years vanadium and some other toxic ions won't be dissociated from TC4 due to a high stability of TC4. And FSP only adding oxygen atoms into the alloy and results in a nano TiO_2 membrane to TC4 surface while none of the toxic elements were added into TC4. However, some researchers have verified that nano TiO_2 has an antibiotic effect. TiO_2 can even help to reduce the risk of being attacked by bacteria and viruses. In the nanoindentation test, the fact that the hardness of stir zone is higher than the base material (TC4) have also been proved. From this property, the conclusion that TC4 with FSP is more stable than TC4 meaning the toxic ions in surface is more difficult to be released can be deduced.

Based on these, the conclusion that TC4 with nano TiO_2 surface has no cytotoxicity is reasonable and the experiment also confirms it.

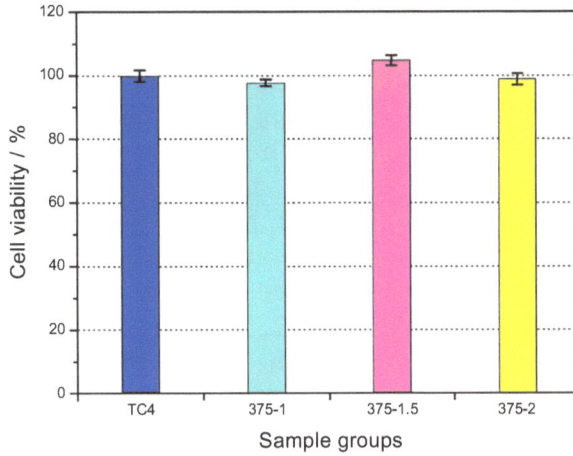

Fig. (6). The cell viability of TC4 with and without FSP in the cytotoxicity test. The bar graph show mean $\pm$ SD [30].

(2) Cell Adhesion Assay

Fig. (7) shows the image of the cell under fluorescence microscope using 200X. Under the microscope cells dyed by DIPI can be seen as blue spots. Each spot corresponds to one cell. Images of different samples are barely different except the number of spots. Using Image-pro plus software to calculate number of blue spots as the number of cells to gain comparisons among all the samples.

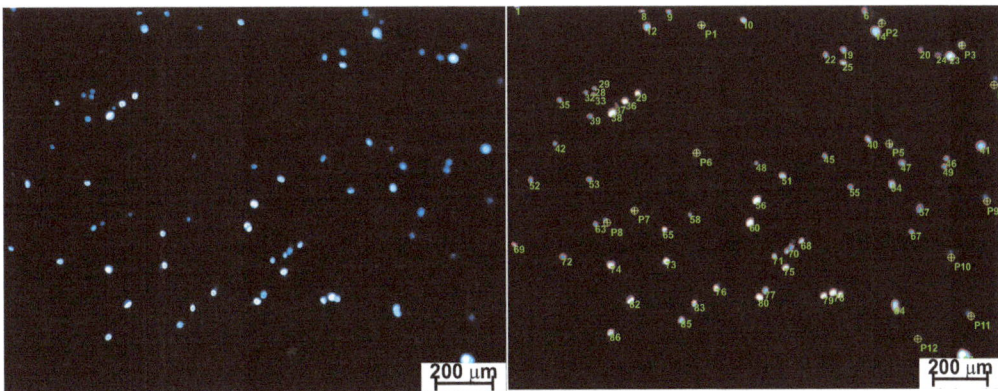

Fig. (7). The cell image on the surface of 375-1 under fluorescence microscope using 200X [30].

As shown in Fig. (8), all of the TC4 processed by FSP groups' cell number is at least 10 percent more than unwrought TC4. This significant difference shows that the nano TiO_2 surface of TC4 made by FSP is more compatible for cell to adhere. There are three advantages of nano TiO_2 surface of TC4 to explain this. First, TiO_2

in nano level can increase the specific surface area obviously, in other words, the contact area between cell and material surface increased. Second, the nano TiO_2 particles can provide many micro-asperities attracting cells to select adhesion position, in other words, cells have more supporting points for themselves to stretch their filiform pseudopodia [40]. Third, surface gaps were open which can make the medium flowing into the gap to support cells. Cells are apt to adhere on this kind of surface. Hence, the adhesion property of FSP groups is better than that of TC4. Fig. (8) also revealed a trend among six groups of TC4 with TiO_2 that the cell number firstly increased then decreased from left to right while 375-1 reached the summit. In three groups of 225, as slot depth increased, there were more TiO_2 to be stirred to increase roughness so that more cells can adhere to the surface. When the rotational speed turned into 375 r/min, the particles stirred were more close to nano scale and more homogeneous which were better for cells adhesion. However, as slot depth increased, the surface becomes too even to adhere more cells than TC4. Hence 375-1 exhibited the best property. Fig. (9) showed the morphology of the cells on the surface of alloy. Compared to TC4, FSP groups had larger spreading areas and better spreading shape as well as more pseudopodia to form a solid adhesion. Cell on TC4 surface was fusiform and had fewer pseudopodia. In contrast, cells on the surface of FSP groups were extended at least one end which made them stretch on the surface more powerful. Since the samples were not polished, some of them occurred scratches which may influence the adhesion of cells. However, all of the groups were disposed under same conditions. And the scratches were smaller than 3µm to which degree the influence can be neglected. Hence, SEM images also demonstrated that FSP made alloy better adhesion behavior.

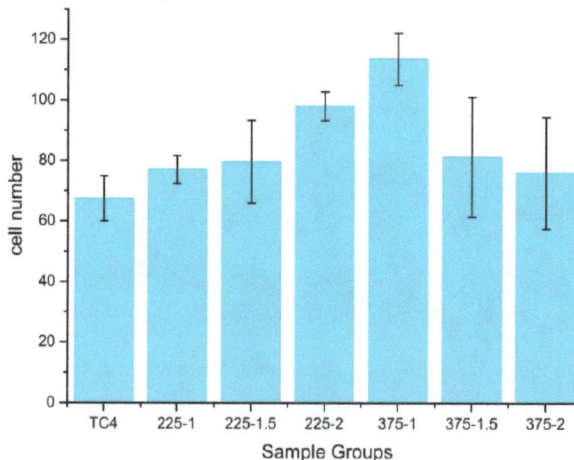

Fig. (8). The cell number of sample groups. Five different areas were pictured randomly from each group and the number was calculated with the help of Image-pro plus software.

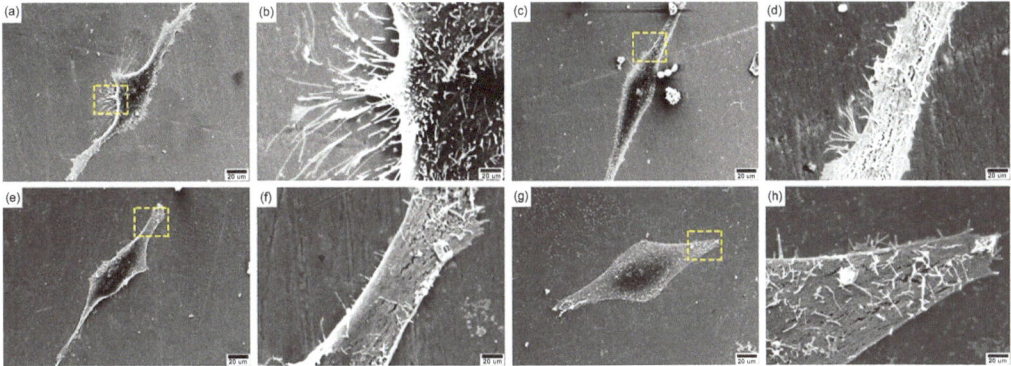

Fig. (9). Mg63 cell in cell adhesion assay under SEM: (**a**) 375-1; (**b**) Enlarged area of the selected area in (a); (**c**) 375-1.5; (**d**) Enlarged area of the selected area in (c); (**e**) 375-2; (**f**) Enlarged area of the selected area in (e); (**g**) TC4 and (**h**) Enlarges area of the selected area in (g) [30].

(3) Cell Proliferation Assay

As shown in Fig. (**10**), in 24h, all of the FSP groups have larger OD than TC4. Although the difference is not obvious, it shows a good trend. However, in 4 and 8 day, there is an abnormal phenomenon that the optical density of 225-1 and 375-2 are much lower than TC4's while other four FSP samples have larger OD than TC4. The most probable reason is that there were some operation mistakes during the experiment. Some cells may be extracted by pipette during medium change after 1 day cell culture. Or some viruses and bacteria were brought into wells accidentally during medium change resulted in the death of some cells. The influence of FSP can be excluded since all the other were in normal trend. So remove 225-1 and 375-2 from the FSP group and use the other four kind of samples as a new FSP group. In the following analysis, consider FSP group has four parameters. Though the FSP group has higher OD, the improvement is not apparent except 375-1 and 375-1.5. The conclusion is that FSP group has a better biocompatibility than TC4 but some of the effect is not obvious. The nano TiO_2 particles created some surface gaps which are not closed and are beneficial for the medium to flow into. That is to say, cells in FSP groups can gain sufficient nutrition which can make cell function better such as proliferation and differentiation. Therefore the nano TiO_2 surface is better for cell proliferation. The 225 r/min is not fast enough to crush all the TiO_2 to nano level. When the slot depth increases, though there are more TiO_2 to be crushed, the number of surface defect was increasing, too. The latter's effect is in an dominant position. Therefore, 225-2 is the best in 225. 375 r/min is fast enough to crush all the TiO_2 in 1mm slot depth. That is, the grain size in 375-1 is the smallest in FSP groups which can provide the best substrate for cells to proliferate. When the slot depth increases, more TiO_2 was added. As a result, the agitation speed can only crush TiO_2 to a size larger than the size of 375-1. The consequence is that the number of surface defect decreases and the length of grain boundary as well as the roughness reduce. Consequently, the 375-1 is the best among three samples processed by 375r/min. In summary, the 375-1 is the best in FSP group.

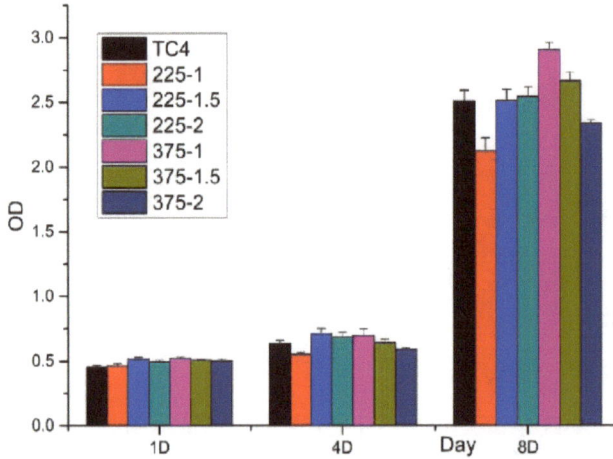

Fig. (10). The optical density of different samples and different cell culture days [30].

Corrosion Behavior

The potentiodynamic polarization studies were carried out on 375-1, 375-1.5, 375-2 and TC4 samples immersed in PBS and Hank's solution. The results are shown in Fig. (**11**). Comparing Tafel plot in PBS and Hank's solution, it can be found that in Hank's solution samples display evident passivated behavior in the potential range between -0.5V and 0V, which means a passive film has formed in Hank's solutions, while in PBS the corrosion current keeps increasing during the same range [41]. In PBS, the corrosion potentials (E_{corr}) are -0.226V for 375-1, -0.414V for 375-1.5, -0.545V for 375-2 and -0.485V for TC4. The corrosion current densities (I_{corr}) are 0.144μA for 375-1, 0.349μA for 375-1.5, 0.078μA for 375-2 and 0.125μA for TC4. In Hank's solution, the corrosion potentials (E_{corr}) are -0.548V for 375-1, -0.570V for 375-1.5, -0.899V for 375-2 and -0.757V for TC4. The corrosion current densities (I_{corr}) is 0.404μA for 375-1, 0.801μA for 375-1.5, 0.453μA for 375-2 and 0.397μA for TC4. All the results have been summarized in Table **1**. The corrosion potentials mean the critical potential beyond which the passive film dissolves and breaks down on the alloy and the value is usually used to describe the stability of the film on the surface of metal. The greater the value is, the more stable the passive film [42]. When amount of TiO_2 added is appropriate (from 1mm to 1.5mm), the corrosion potential of samples would increase compared with TC4, which means the possibility of corrosion reduces. However, in this range (1.5-2mm), the amount of TiO_2 added has a negative proportional relationship with corrosion potential. Since the only variable quantity is the amount of TiO_2, it can be concluded that FSP can help increase the corrosion potential while adding TiO_2 decreases the corrosion potential. But in a reasonable range, considering the effect from both FSP and adding TiO_2, the method can be benefit to increase the corrosion potential. The difference of data between two solutions may result from difference ions in two solutions (as Hank's solution has extra HCO_3^-, Ca^{2+}, Mg^{2+} and SO_4^{2-}). When it comes to corrosion current density, no exact relationship between amount of TiO_2 and I_{corr} can be concluded. Except data of 375-2 in PBS, the corrosion current density of all samples increases with the slight increment extent.

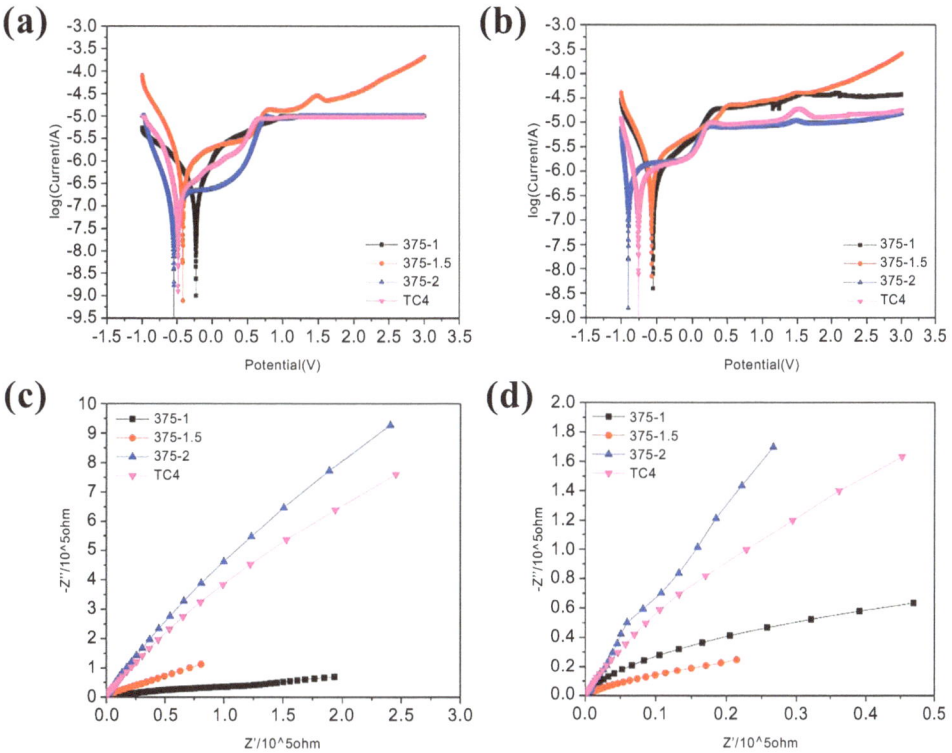

Fig. (11). Potentiodynamic polarization curves samples in (**a**) PBS and (**b**) Hank's solution; electrochemical impedance spectra of samples in (**c**) PBS and (**d**) Hank's solution [30].

Fig. (**9c-d**) show Nyquist plot of samples in two solutions. In the figure, the calculation of Z' and Z'' can refer to the equation:

$$Z' = \frac{R_p}{1+(\omega R_p C_d)^2} \tag{1}$$

$$Z'' = \frac{\omega R_p^2 C_d}{1 + (\omega R_p C_d)^2} \tag{2}$$

Impedance of material can be obtained according to the curvature (or radius) of the curve [43]. From the Figures, 375-2 has the highest impedance among all samples subjected to FSP. In both solutions, 375-2 has higher impedance than TC4. All other samples have lower impedance and no specific relationship can be concluded between the amount of TiO_2 and the resistance of materials.

CONCLUSION

According to observation and comparison of metallograph of TC4 after FSP processing, we can find that when feeding speed is fixed and mixing speed is not very high, FSP processing will obviously refine crystalline grains of surface. After FSP processing, crystalline grains will refine and break.

There are three different zones after FSP processing, which are stirring zone (mechanically affected zone), heat-engine-affected zone and heat-affected zone. As the last two zones are difficult to distinguish, during experiment we regard them as transition zone. In stirring zone (mechanically affected zone) material is mainly affected by mechanical mixing of mixing head therefore will break and refine significantly. Below temperature of phase transformation the higher speed mixing head has, the better crystalline grains will refine. As heat-affected-zone is far away from mixing head, refinement in this zone can be neglected. So the main variation tendency is growth of crystalline grains because of friction heat thus size of crystalline grains heat-affected-zone is bigger than that of base metal.

Among three different zones, using nanoindentation, we have verified that the hardness of stir zone is the most among three zones while the base material is the second and transition zone is the third. FSP has crushed the particles to nano-sized successfully and increased the harness on the surface. This also demonstrates the wear property has also been promoted meaning less ions or particles getting into adjacent area caused by abrasion. This improvement can alleviate the problem of prosthesis' damage to human body.

Based on the three vitro biological assays, we can primarily prove that TC4 after FSP processing has a better biocompatibility than TC4. At first, the new material has no cytotoxicity as TC4. Second, the surface of TC4 with nano TiO_2 composite layer is more suitable for cells to adhere. And the adhesion of cells is the fundamental and necessary process that is of significance to the following proliferation and differentiation. Third, TC4 processed by FSP is better for cells' proliferation than TC4 which can make the bone union faster and is beneficial to the combination of prosthesis and bone. In summary, TC4-TiO$_2$ composite material is excellent for biomedical use.

CONFLICT OF INTEREST

The author (editor) declares no conflict of interest, financial or otherwise.

ACKNOWLEDGEMENTS

Declared None

REFERENCES

[1]   Okazaki Y, Rao S, Ito Y, Tateishi T. Corrosion resistance, mechanical properties, corrosion fatigue strength and cytocompatibility of new Ti alloys without Al and V. Biomaterials 1998; 19(13): 1197-215.
[http://dx.doi.org/10.1016/S0142-9612(97)00235-4] [PMID: 9720903]

[2]   Eisenbarth E, Velten D, Müller M, Thull R, Breme J. Biocompatibility of beta-stabilizing elements of titanium alloys. Biomaterials 2004; 25(26): 5705-13.
[http://dx.doi.org/10.1016/j.biomaterials.2004.01.021] [PMID: 15147816]

[3]   Leyens C, Peters M. Titanium and titanium alloys: fundamentals and applications 2006.

[4]   Polmear IJ. Light alloys: from traditional alloys to nanocrystals. Elsevier 2006.

[5]   Chan KS, Koike M, Okabe T. Modeling wear of cast Ti alloys. Acta Biomater 2007; 3(3): 383-9.
[http://dx.doi.org/10.1016/j.actbio.2006.10.007] [PMID: 17224314]

[6]   Teoh SH. Fatigue of biomaterials: a review. Int J Fatigue 2000; 22: 825-37.
[http://dx.doi.org/10.1016/S0142-1123(00)00052-9]

[7]   Rack HJ, Qazi JI. Titanium alloys for biomedical applications. Mater Sci Eng C 2006; 26: 1269-77.
[http://dx.doi.org/10.1016/j.msec.2005.08.032]

[8]   Jiang YL. Study of mechanism of electrochemical reaction for titanium alloy TC4 in 3% NaCl solution and ethanol by polarization curve. Corros Sci Prot Technol 2005.

[9]   Mishra RS, Mahoney MW, McFadden SX, Mara NA, Mukherjee AK. High strain rate superplasticity in a friction stir processed 7075 Al alloy. Scr Mater 1999; 42: 163-8.
[http://dx.doi.org/10.1016/S1359-6462(99)00329-2]

[10] Mishra RS, Mahoney MW. Friction stir processing: a new grain refinement technique to achieve high strain rate superplasticity in commercial alloys. Mater Sci Forum 2001; 357-359(3): 507-14.
[http://dx.doi.org/10.4028/www.scientific.net/MSF.357-359.507]

[11] Mishra RS, Ma ZY. Friction stir welding and processing. Mater Sci Eng Rep 2005; 50: 1-78.
[http://dx.doi.org/10.1016/j.mser.2005.07.001]

[12] Ma ZY. Friction stir processing technology: a review. Metall Mater Trans A 2008; 39: 642-58.
[http://dx.doi.org/10.1007/s11661-007-9459-0]

[13] Jialei WU, Wang K, Zhou L, Wang W. Development of friction stir processing. Hot Working Technol 2010; 39: 150-3.

[14] Huang C, Ke L, Li X, Liu G. Research progress and prospect of friction stir processing. Rare Met Mater Eng 2011; 40: 183-8.

[15] Huiskes R, Ruimerman R, van Lenthe GH, Janssen JD. Effects of mechanical forces on maintenance and adaptation of form in trabecular bone. Nature 2000; 405(6787): 704-6.

[http://dx.doi.org/10.1038/35015116] [PMID: 10864330]

[16]   Hangai Y, Utsunomiya T, Hasegawa M. Effect of tool rotating rate on foaming properties of porous aluminum fabricated by using friction stir processing. J Mater Process Technol 2010; 210: 288-92.
[http://dx.doi.org/10.1016/j.jmatprotec.2009.09.012]

[17]   Feng AH, Ma ZY. Enhanced mechanical properties of Mg-Al-Zn cast alloy *via* friction stir processing. Scr Mater 2007; 56: 397-400.
[http://dx.doi.org/10.1016/j.scriptamat.2006.10.035]

[18]   Hinz F. Flexible control of the microstructure and mechanical properties of friction stir welded Ti-6A--4V joints. Mater Des 2013; 46: 348-54.
[http://dx.doi.org/10.1016/j.matdes.2012.10.051]

[19]   Li B, Shen Y, Hu W, Luo L. Surface modification of Ti-6Al-4V alloy *via* friction-stir processing: microstructure evolution and dry sliding wear performance. Surf Coat Tech 2014; 239: 160-70.
[http://dx.doi.org/10.1016/j.surfcoat.2013.11.035]

[20]   Ma H, Wang K. Microstructure and properties of TA2 commercially pure titanium surface by friction stir processing. Rare Met Mater Eng 2012; 40: 1530-3.

[21]   Roy SC, Paulose M, Grimes CA. The effect of TiO_2 nanotubes in the enhancement of blood clotting for the control of hemorrhage. Biomaterials 2007; 28(31): 4667-72.
[http://dx.doi.org/10.1016/j.biomaterials.2007.07.045] [PMID: 17692372]

[22]   Eriksson C, Lausmaa J, Nygren H. Interactions between human whole blood and modified TiO_2-surfaces: influence of surface topography and oxide thickness on leukocyte adhesion and activation. Biomaterials 2001; 22(14): 1987-96.
[http://dx.doi.org/10.1016/S0142-9612(00)00382-3] [PMID: 11426876]

[23]   Yang B, Uchida M, Kim HM, Zhang X, Kokubo T. Preparation of bioactive titanium metal *via* anodic oxidation treatment. Biomaterials 2004; 25(6): 1003-10.
[http://dx.doi.org/10.1016/S0142-9612(03)00626-4] [PMID: 14615165]

[24]   Keller TF, Reichert J, Thanh TP, *et al.* Facets of protein assembly on nanostructured titanium oxide surfaces. Acta Biomater 2013; 9(3): 5810-20.
[http://dx.doi.org/10.1016/j.actbio.2012.10.045] [PMID: 23142481]

[25]   Liu X, Chu PK, Ding C. Surface modification of titanium, titanium alloys, and related materials for biomedical applications. Mater Sci Eng Rep 2004; 47: 49-121.
[http://dx.doi.org/10.1016/j.mser.2004.11.001]

[26]   Mello MG, Taipina MO, Rabelo G, Cremasco A, Caram R. Production and characterization of TiO_2 nanotubes on Ti-Nb-Mo-Sn system for biomedical applications. Surf Coat Tech 2017; 326: 126-33.
[http://dx.doi.org/10.1016/j.surfcoat.2017.07.027]

[27]   Huang Q, Liu X, Elkhooly TA, *et al.* Preparation and characterization of TiO_2/silicate hierarchical coating on titanium surface for biomedical applications. Mater Sci Eng C 2016; 60: 308-16.
[http://dx.doi.org/10.1016/j.msec.2015.11.056] [PMID: 26706535]

[28]   Araghi A, Hadianfard MJ. Fabrication and characterization of functionally graded hydroxyapatite/TiO_2 multilayer coating on Ti-6Al-4V titanium alloy for biomedical applications. Ceram Int 2015; 41: 12668-79.
[http://dx.doi.org/10.1016/j.ceramint.2015.06.098]

[29]   Rosales-Rocabado JM, Kaku M, Kitami M, Akiba Y, Uoshima K. Osteoblastic differentiation and mineralization ability of periosteum-derived cells compared with bone marrow and calvaria-derived cells. J Oral Maxillofac Surg 2014; 72(4): 694.e1.
[http://dx.doi.org/10.1016/j.joms.2013.12.001] [PMID: 24480775]

[30]   Zhang CJ, Ding ZH, Xie LC, *et al.* Electrochemical and *in vitro* behavior of the nanosized composites of Ti-6Al-4V and TiO_2 fabricated by friction stir process. Appl Surf Sci 2017; 423: 331-9.
[http://dx.doi.org/10.1016/j.apsusc.2017.06.141]

[31]   Ding ZH, Zhang CJ, Xie LC, Zhang LC, Wang LQ, Lu WJ. Effects of friction stir processing on the phase transformation and microstructure of TiO_2 compounded Ti-6Al-4V alloy. Metall Mater Trans, A Phys Metall Mater Sci 2016; 47: 5675-9.
[http://dx.doi.org/10.1007/s11661-016-3809-8]

[32]   Li B, Shen YF, Hu WY, Luo L. Surface modification of Ti-6Al-4V alloy *via* friction-stir processing: microstructure evolution and dry sliding wear performance. Surf Coat Tech 2014; 239: 160-70.
[http://dx.doi.org/10.1016/j.surfcoat.2013.11.035]

[33]   Zhang X, Wang Q, Mo W. Physical metallurgy and heat treatment of titanium. Metall Ind Press 2009.

[34]   Wang L, Qu J, Chen L, *et al.* Investigation of Deformation Mechanisms in β-Type Ti-35Nb-2Ta-3Zr Alloy *via* FSP Leading to Surface Strengthening. Metall Mater Trans A 2015; 46: 4813-8.
[http://dx.doi.org/10.1007/s11661-015-3089-8]

[35]   Zhang X, Zhao Y, Bai C. Titanium alloy and its application. Beijing: Chem Ind Press 2005.

[36]   Liu HJ, Li Z. Microstructural zones and tensile characteristics of friction stir welded joint of TC4 titanium alloy. T Nonferr Metal Soc 2010; 20: 1873-8.
[http://dx.doi.org/10.1016/S1003-6326(09)60388-5]

[37]   Oliver WC, Pharr GM. Measurement of hardness and elastic modulus by instrumented indentation: advances in understanding and refinements to methodology. J Mater Res 2004; 19: 3-20.
[http://dx.doi.org/10.1557/jmr.2004.19.1.3]

[38]   Oak JJ, Inoue A. Attempt to develop Ti-based amorphous alloys for biomaterials. Mater Sci Eng A 2007; 449: 220-4.
[http://dx.doi.org/10.1016/j.msea.2006.02.307]

[39]   de Lima GG, de Souza GB, Lepienski CM, Kuromoto NK. Mechanical properties of anodic titanium films containing ions of Ca and P submitted to heat and hydrothermal treatment. J Mech Behav Biomed Mater 2016; 64: 18-30.
[http://dx.doi.org/10.1016/j.jmbbm.2016.07.019] [PMID: 27479891]

[40]   Takeda I, Kawanabe M, Kaneko A. An investigation of cell adhesion and growth on micro/nano-scale structured surface-Self-assembled micro particles as a scaffold. Precis Eng 2016; 43: 294-8.
[http://dx.doi.org/10.1016/j.precisioneng.2015.08.009]

[41]   Dai N, Zhang LC, Zhang J, Chen Q, Wu M. Corrosion behavior of selective laser melted Ti-6Al-4 V alloy in NaCl solution. Corros Sci 2016; 102: 484-9.
[http://dx.doi.org/10.1016/j.corsci.2015.10.041]

[42]   Dai N, Zhang LC, Zhang J, *et al.* Distinction in corrosion resistance of selective laser melted Ti-6Al-4V alloy on different planes. Corros Sci 2016; 111: 703-10.
[http://dx.doi.org/10.1016/j.corsci.2016.06.009]

[43]   Goldman DE. Potential, impedance, and rectification in membranes. J Gen Physiol 1943; 27(1): 37-60.
[http://dx.doi.org/10.1085/jgp.27.1.37] [PMID: 19873371]

Corrosion Behavior of Selective Laser Melted Titanium Alloys

Lai-Chang Zhang[1,*], Junxi Zhang[2] and Liqiang Wang[3]

[1] *School of Engineering, Edith Cowan University, Perth, WA, Australia*

[2] *Shanghai Key Laboratory of Material Protection and Advanced Material in Electric Power, Shanghai University of Electric Power, Shanghai, China*

[3] *State Key Laboratory of Metal Matrix Composites, Shanghai Jiao Tong University, Shanghai, China*

Abstract: It is well known that different manufacturing technologies of titanium alloys have a substantial impact on its performance; for example the Ti-6Al-4V manufactured by selective laser melting (SLM) exhibit comparable even better mechanical properties than the counterpart produced by traditional manufacturing methods. Yet, the understanding of the corrosion behavior of SLM-produced materials is unknown. This chapter reviews the recent progress of the corrosion behavior of SLM-produced Ti-based alloys, such as Ti-6Al-4V and Ti-TiB composite. The corrosion behavior and corrosion mechanism are compared and discussed between the SLM-produced titanium alloys and their counterparts processed by traditional methods.

Keywords: Composite, Corrosion behavior, Electrochemical methods, Microstructure, Selective laser melting, Ti-6Al-4V.

INTRODUCTION

It is well known that the dense and chemically stable oxide film formed spontaneously on the surface of a metal is the fundamental cause for high corrosion resistance [1 - 4]. Titanium alloys are widely used in the marine, automotive and aerospace industries because of their endothermic properties, such as high strength to weight ratio, super-plasticity, excellent biocompatibility and formability [1, 5 - 11]. Ti-6Al-4V is the most commonly used titanium alloy for engineering parts, biomedical and dental implants [12 - 14]. Similarly, Ti-6Al-4V based on its wide range of microstructure, including hexagonal close packed (hcp) α phase and body-centered cubic (bcc) β phase, can also be considered as heat-

---
*Corresponding author Lai-Chang Zhang:** School of Engineering, Edith Cowan University, 270 Joondalup Drive, Joondalup, Perth, WA 6027, Australia; Tel: 61 8 63042322; Fax: 61 8 63045811; E-mails: lczhangimr@gmail.com; l.zhang@ecu.edu.au

treatable two-phase structural titanium alloy. Furthermore, the α/β volume ratio and the chemical composition of each constituent phase can be adjusted for optimizing its properties according to the heat treatment [3]. Considerable endeavors have been made to study the corrosion behavior of titanium alloys in harsh conditions and various media simulations of the human body [15 - 22].

Traditionally, titanium and its alloys are manufactured by the processing technologies such as casting, machining, sheet forming and powder metallurgy [23 - 26]. However, these traditional manufacturing methods used to produced Ti-based materials profoundly involve time, energy and materials consumption [27], resulting in high cost of titanium produces. Therefore, the development of new and accessible technologies that involve relatively fast production time and less material waste is needed, such as additive manufacturing (called 3D printing technique) [23, 27]. Selective laser melting (SLM) technology is one of the most frequently used additive manufacturing techniques to produce titanium based composites because this method is computer controlled, using a laser to generate higher energy and melt the metal power instantaneously under the protective atmosphere [27 - 30]. Due to these features, there exist a significant number of promising applications in the production of replacement parts, dental crowns, and artificial limbs, as well as bridge manufacturing. Furthermore, the SLM-produced titanium parts show comparable or even much better mechanical properties [23, 27, 31 - 34]. For example, the SLM-produced β-type Ti-24Nb-4Zr-8Sn components have comparable mechanical properties to those prepared through traditional manufacturing methods [23, 28, 30]. The SLM-processed CP-Ti samples show better wear resistance than their cast counterparts but with similar wear mechanisms [31]. According to these results, it seems that the alloys produced by SLM method could have some comparable or even better properties than traditional techniques. As expected, compared with Ti-6Al-4V processed by conventional methods, the SLM produced Ti-6Al-4V usually shows improved mechanical properties. This is mainly attributed to the fine microstructure formed from rapid solidification in the SLM manufacturing process [10]. However, the microstructural characterizations indicate that the SLM-produced Ti-6Al-4V is dominated by fine acicular α′ martensite together with some prior β grains [35 - 37], which is prominently different from the typically α+β biphasic microstructure in traditionally produced Grade 5 alloy. It is known that overpowering acicular α′ martensite exists in the microstructure of SLM-produced Ti-6Al-4V alloys, especially along with long columnar β grains. At the same time, some pores and melt pools also co-exist in the microstructure. It is reported that these effects are mainly determined by the SLM scanning parameters [34].

Because the microstructure of the titanium alloy manufactured by SLM is different from that of the conventional method (as proved by [27, 31 - 33]), these hierarchical manufacturing characteristics of the SLM will correspond to some differences of mechanical properties as well as corrosion resistance. Furthermore, the corrosion resistance of the titanium alloy made by SLM technique may also be related to the difference between the sample planes (*i.e.* the build plane and the growth plane). Unfortunately, a large number of needle-like α' martensitic phases present in the Ti-6Al-4V alloy made by SLM technique makes it less resistant to corrosion than the Ti-6Al-4V alloy manufactured by conventional methods. The alloys made by the SLM technique often exhibits a stressed, un-stabilized and segregated microstructure due to its specific scanning strategies [35]. Also, the XZ-plane (*i.e.* the growth plane) of the SLM-produced Ti-6Al-4V alloy is composed of more α'-Ti and less β-Ti phase than XY-plane (*i.e.* the build plane), which leads to the inferior corrosion resistance of XZ-plane [38]. Recently, Bai *et al.* [39] found that the Ti-6Al-4V alloy fabricated by electron beam melting includes α phase rather than fine α' martensite, and improved the corrosion resistance compared to conventional Ti-6Al-4V alloy (Grade 5). Therefore, it is expected that the corrosion resistance of Ti-6Al-4V produced by SLM can be improved if the non-acicular α' martensite phase can be transformed into α martensitic phase by heat treatment. However, there have been some reports on the effects of heat treatment on the microstructure of SLM-produced Ti-6Al-4V alloys [36, 40 - 42]. Dai *et al.* [43] also carried out heat treatments on the SLM-produced Ti-6Al-4V at 500°C, 850°C and 1000°C, respectively. But their results are not ideal. The corrosion resistance is reduced with increasing the heat treatment temperature, even the passive film is formed on the surface of samples that have undergone the heat treatment of 1000°C. So, in the context of this chapter, the discussion will elucidate some aspects of the corrosion resistance by using electrochemical methods. This chapter focuses on the comparison of the corrosion behaviors between different manufacturing methods and planes, or after various heat treatments, and other titanium composites.

CORROSION BEHAVIOR OF SLM-PRODUCED TI-6AL-4V

It is well known that the electrode potential of the metal surface substantially reflects the surface state of the sample when it is immersed in the electrolyte. As shown in Fig. (**1a**), the potential of the Ti-6Al-4V alloy shifts forward in the NaCl solution medium. It is clear that the Ti-6Al-4V alloy is naturally inhibited by the formation of the passivation film under the electrolyte. For Grade 5 samples, Open Circuit Potential (OCP) reaches a stabilization state at 25 hours. By contrast, the alloy made by SLM is still slightly increased even after 60 hours. The SLM-produced Ti-6Al-4V alloy appears to have a slightly higher final stabilization potential than the commercial grade 5 alloy, compared with its relatively stable OCP. Nevertheless, there is almost no difference in the corrosion resistance of Ti-6Al-4V made by SLM in different planes under NaCl solution.

Therefore, Fig. (**1b**) shows the OCP curve with a harsher corrosion environment (1M HCl solution) of the XY and XZ plane of SLM produced Ti-6Al-4V alloy, respectively. Similar to Fig. (**1a**), the OCP values of the two planes remain forward offset when immersed in HCl due to the formation of the passivation film on the surface of the alloy. After 45 h of immersion, it is difficult for the XY and XZ planes to obtain a completely stable potential. But the potential fluctuation is less than 1.5 mV/h, and this kind of slight change in potential has no significant effect on the subsequent Electrochemical Impedance Spectroscopy (EIS) test.

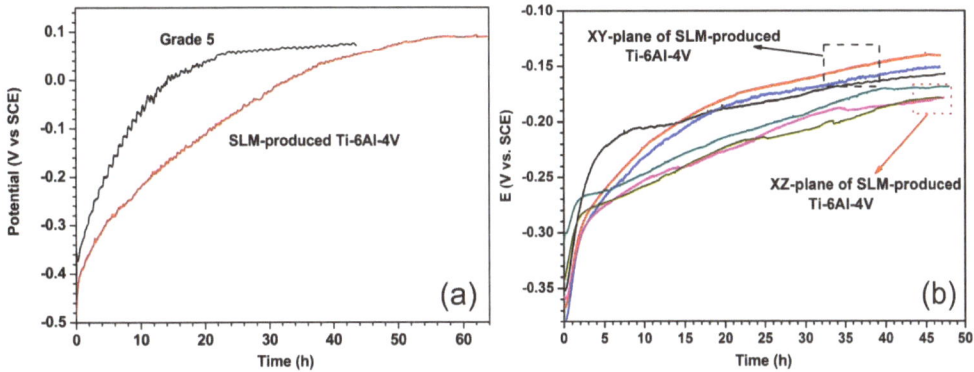

Fig. (1). The open circuit potential of (**a**) commercial Grade 5 alloy *vs.* the SLM-produced Ti-6Al-4V alloy, (**b**) the SLM-produced Ti-6Al-4V alloy with different planes [35, 38].

Fig. (**2**) shows the potentiodynamic polarization curves of Ti-6Al-4V alloy produced by SLM and commercial grade 5 alloy in NaCl solution. It can be seen that these polarization curves have very good reproducibility. Furthermore, the Tafel slope cannot be clearly defined in the anode region. Such a phenomenon indicates that the Ti-6Al-4V alloy prepared by the SLM technique and the conventional technique exhibits a corrosion behavior. The $i_{P,A}$ in Fig. (**2**) represents the passivation current of the commercial grade 5 alloy and has a value of 0.390 ± 0.0125 $\mu A/cm^2$. The passivation current value ($i_{P,B}$) of Ti-6Al-4V alloy produced by SLM is about two times that of the commercial grade 5 samples (0.841 ± 0.0275 $\mu A/cm^2$).

The points $E_{b,A}$ and $E_{b,B,}$ respectively, represent the critical potential of the passivation film to dissolve and decompose on the commercial Grade 5 alloy and the SLM-produced Ti-6Al-4V alloy. The potential (E_b) at this point is usually used to describe the stability of the film on the metal surface. In general, the larger the value of E_b, the more stable the oxide film. It is easy to observe from Fig. (**2**) that $E_{b,A}$ is greater than $E_{b,B}$. This result indicates that the passivation film on the surface of the commercial grade 5 alloy sample is more stable than the passivation film formed on the SLM-produced Ti-6Al-4V alloy sample surface. So, as observed from the potentiodynamic polarization curves, the corrosion resistance of commercial grade 5 alloy is better than the SLM-produced sample when the sample immersed in 3.5 wt.% NaCl solution.

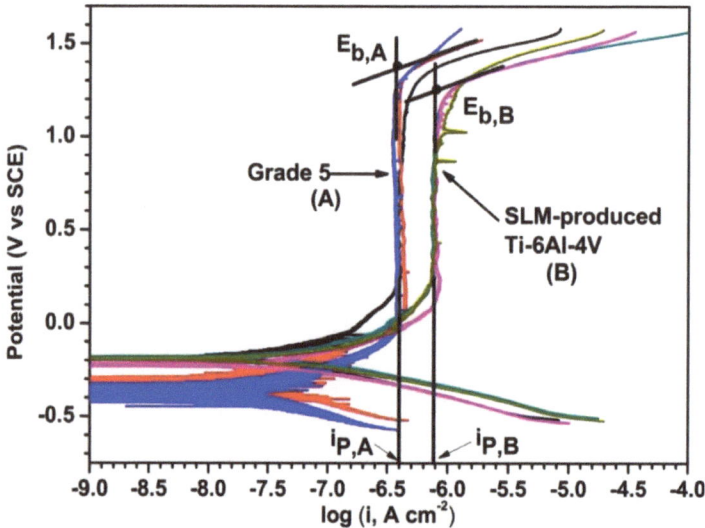

Fig. (2). Potentiodynamic polarization curves for the SLM-produced Ti-6Al-4V alloy and commercial Grade 5 alloy in NaCl solution [35].

Fig. (**3**) shows the potentiodynamic polarization curves of the SLM produced Ti-6Al-4V alloy in XY plane and XZ plane, and immersed in 3.5 wt.% NaCl solution and 1M HCl solution, separately. On the XY plane and XZ plane, the passivation behavior of Ti-6Al-4V produced by SLM in 1M HCl solution was between 200 mV and 1500 mV, and the alloy immersed at 3.5wt. % NaCl solution is in the range of 200 mV to 1200 mV (the potential range is less than immersed in 1M HCl solution). This means that the alloy in the corrosive solution forms a passivation film within the above-mentioned potential range, which serves as a protective film to suppress the corrosion of the titanium alloy in the corresponding solution medium. It can be seen from the Fig. (**3**) that i_p represents the passivation current of the film formed on the surface of the alloy, and usually the lower i_p indicates that the alloy is susceptible to passivation. Thus, the lower readily passivated Ti-6Al-4V alloy generally exhibits better corrosion resistance in the corresponding solution system. So, the corrosion behavior of Ti-6Al-4V produced by SLM in the NaCl solution is very close to that of the XZ plane. However, in the harsher condition of 1M HCl solution, $i_{P,XY}$ and $i_{P,XZ}$ are referred to the passivation current of XY-planes and XZ-planes, respectively. The value of the passivation current for the XZ-plane ($i_{P,XZ}$) of SLM-produced Ti-6Al-4V alloy is

2.83 ± 0.04 μA·cm^{-2}, and the $i_{P,XY}$ is 2.50 ± 0.02 μA·cm^{-2}, which is slightly lower than the $i_{P,XZ}$. The smaller value of i_P also indicates a slow dissolution of the alloy. Therefore, the XY plane alloy of Ti-6Al-4V produced by SLM has a slightly better corrosion resistance than the XZ plane in the 1M HCl solution. Hence, the SLM produced Ti-6Al-4V exhibits a higher corrosion rate in 1M HCl solution compared to 3.5 wt% NaCl solution. At the same time, the difference in corrosion resistance is more prominent in XY-planes than XZ-planes in 1M HCl solution.

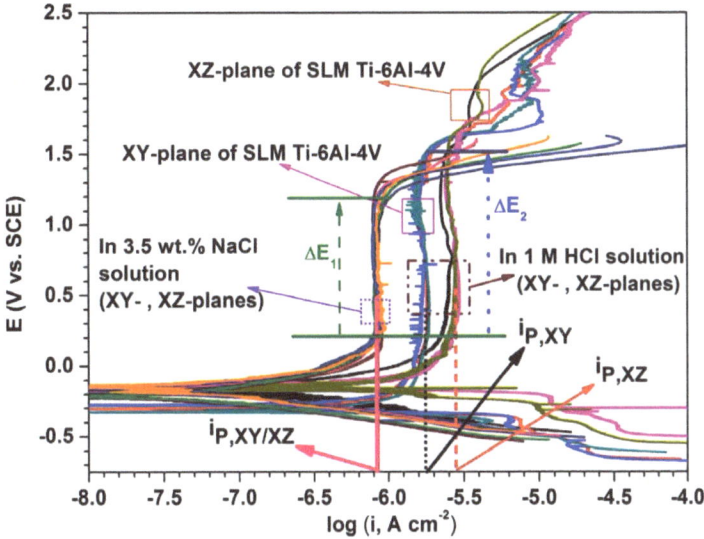

Fig. (3). Potentiodynamic polarization curves for the XY and XZ-planes of SLM-produced Ti-6Al-4V alloy in 3.5 wt.% NaCl solution and 1 M HCl solution [38].

The EIS measurements are usually shown in the form of Nyquist and Bode graphs, as shown in Fig. (**4a-c**). The EIS data, analyzed using an equivalent circuit with two-time constants, were used to compare the corrosion behavior between the SLM produced Ti-6Al-4V and commercial grade 5. The equivalent circuit (Fig. **4d**) consists of solution resistance (R_s), charge transfer resistance (R_{ct}), film resistance (R_f) and constant phase elements (CPE_{dl} and CPE_l). The use of CPE is because the electrode surface is not ideally flat [6]. The first two rows of Table **1** list the fitting results of the EIS measurements. It can be seen from the fitting results that the value of Ti-6Al-4V alloy produced by SLM are smaller than the result of commercial grade 5 alloy. Regarding the film resistance (R_f), the larger the value of R_f, the better the corrosion resistance. As a result, the commercial grade 5 alloy has better corrosion resistance than the SLM-produced Ti-6Al-4V composite. Also, the results of EIS correspond to the results obtained from the previous measurement of the passive potential polarization curve. Fig. (**5**) shows the EIS measurements of the XY and XZ planes of the Ti-6Al-4V alloy produced by SLM in 1M HCl solution. The illustration in Fig. (**5**) shows the EIS results of SLM produced Ti-6Al-4V immersed in 3.5 w.t.% NaCl solution. The elements in the equivalent circuit are identical to those in Fig. (**4**). As shown in Fig. (**5a**), the Nyquist graph shows only one large capacitive loop, but the Bode diagram (Fig. **5b**) shows a wide frequency range (from high frequency to low frequency). Thus, an equivalent circuit with two time constants is used to accommodate this EIS data. Table **1** (except the first two lines) summarizes the results of the EIS measurements. For the Ti-6Al-4V alloy produced by SLM in 3.5 w.t.% NaCl solution, the fitting results of R_f (Table **1**) have very close values for the XY plane and the XZ plane; the same as the R_{ct} value. This indicates that there is no significant difference in the corrosion resistance of the XY and XZ planes in 3.5 wt% NaCl solution. In contrast, the R_f value of the XY plane is greater than the R_f value of the XZ plane for the Ti-6Al-4V alloy in the 1M HCl solution (harsher conditions). This reveals that the passivation film formed on the XY plane of the SLM produced Ti-6Al-4V has better protection than the XZ plane. The R_{ct} results also support the conclusion of R_f analysis.

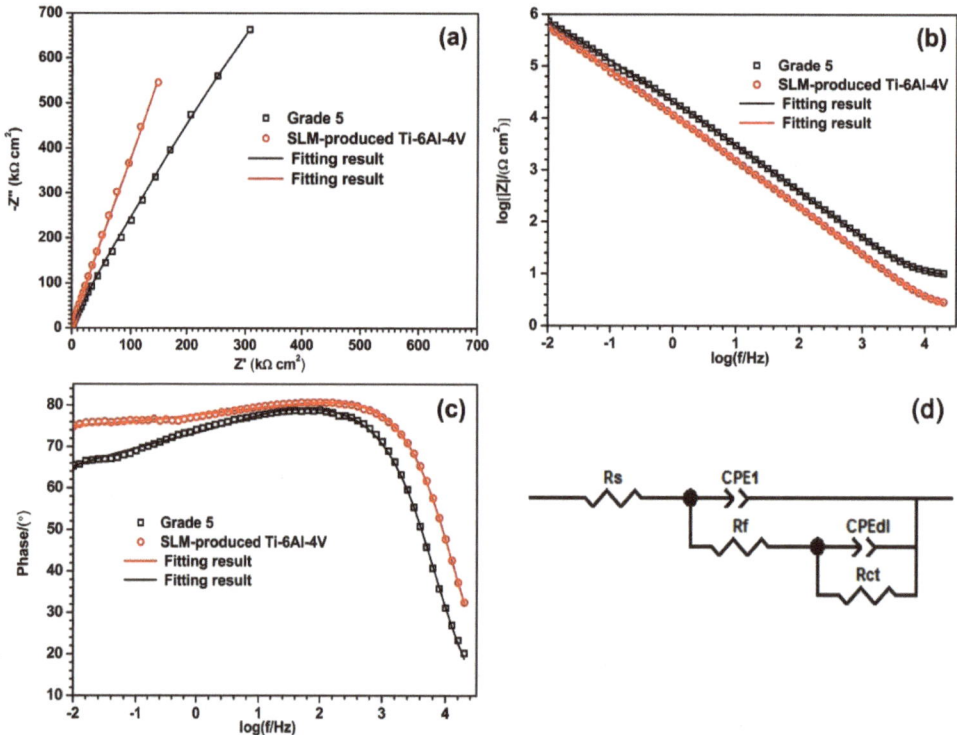

Fig. (4). The EIS measurements: (**a**) Nyquist plot, (**b, c**) Bode plots, and (**d**) equivalent circuit for the SLM-produced Ti-6Al-4V alloy and commercial Grade 5 alloy in NaCl solution [35].

Table 1. The fitting parameters of EIS for commercial Grade 5 alloy and SLM-produced Ti-6Al-4V alloy with different planes.

Solution	Sample	R_f (k$\Omega \cdot$cm^2)	R_{ct} (k$\Omega \cdot$cm^2)	*Ref.*
3.5 wt.% NaCl	SLM-produced, XY-plane	40.26	17410	[35]
	Commercial Grade 5	94.12	58990	[35]
	SLM-produced, XY-plane	26.84	814.90	[38]
	SLM-produced, XZ-plane	22.69	798.30	[38]
1 M HCl	SLM-produced, XY-plane	5.63	293.30	[38]
	SLM-produced, XZ-plane	1.67	129.70	[38]

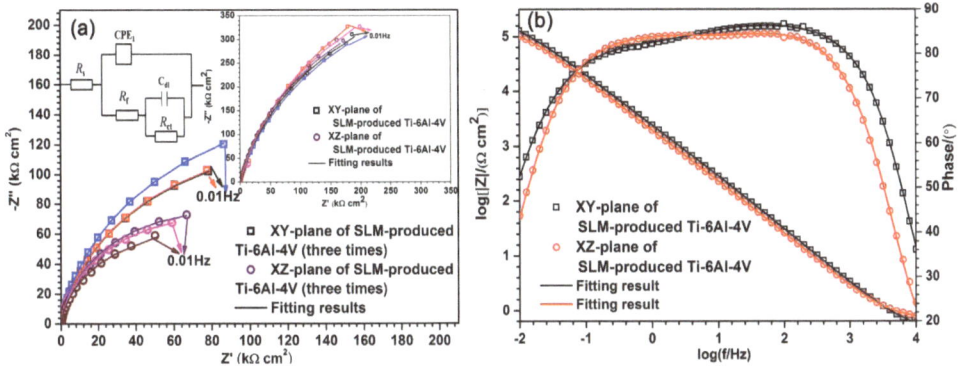

Fig. (5). The EIS measurements for the XY and XZ- planes of SLM-produced Ti-6Al-4V alloy in 1M HCl: **(a)** Nyquist plots, **(b)** Bode plots. The inset figures in **(a)** were Nyquist plots for XY and XZ planes of SLM-produced Ti-6Al-4V alloy in 3.5 wt.% NaCl solution and equivalent circuit used to impedance spectra analysis [38].

Fig. (**6**) shows the optical microstructures of Grade 5 alloy. For commercial grade 5 alloys, both α and β phases can be observed, where the large light color areas and tiny dark color areas are representing α and β phases, respectively. Similar metallographic structures have been obtained in other literature [44]. On the other hand, unlike the commercial grade 5 alloy, the microstructure of the Ti-6Al-4V alloy produced by the SLM does not show a significant α and β phases, but it is replaced by distinct layer boundaries, as shown in Fig. (**7**). The boundaries of the building layers are significant, which are also common and are caused by the hierarchical characteristics of the SLM technique [23]. Considering these hierarchical characteristics, it is also reasonable to observe only the layers in the microstructure [23]. The microstructures and mechanical properties of Ti-6Al-4V produced by SLM method have also been studied in some literature [36, 37, 45, 46]. According to the nature of SLM process, the powder is heated extremely fast and the produced solid layer is also cooled at a high rate, thereby usually resulting in the formation of a non-equilibrium phase or microstructure [27, 34], such as martensitic phase. This non-equilibrium microstructure will be deformed and may be in a "higher energy state" concerning corrosion. It can also be observed in Fig. (7) that some long, columnar grains are observed in the building direction, more or less. In the SLM process, there is no nucleation of the nucleation barrier, which leads to the formation of elongated grains [37]. These grains are considered to be epitaxially grown in the SLM process, and the length can be up to several millimeters [37].

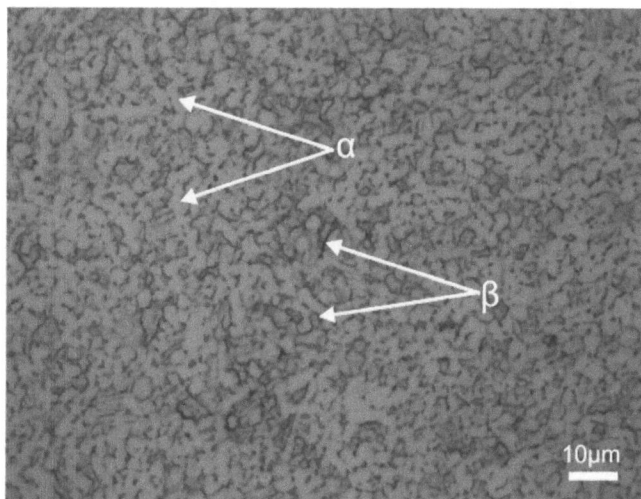

Fig. (6). Optical microstructure of Grade 5 alloy [35].

Overall, the XZ plane and the XY plane of the Ti-6Al-4V alloy for SLM production show a quantitative needle-like α' martensite, which is the clear majority phases in the optical micrograph of Fig. (7). There are also some defects, such as several pores are investigated in the microstructure of the XY plane and the XZ plane. Pores are easily elongated along the scanning direction and are located in certain layers. The rupture accumulation and interlayer surface roughness around the pool in the building layer is considered to be the main cause of pore formation [37]. Different from the cylinder β grains in the XZ plane (Fig. 7a), the primary β grains in the XY plane are square, and the scanning strategy used in the process of SLM is close to each other. The occurrence of acicular α' martensite is attributed to the followings. (i) a significantly higher cooling rate (103-108°C/s) in the SLM process and the transformation of the martensite through β in Ti-6Al-4V (About 410 K/s) of α', and/or (ii) the direction of construction in the micro-bath along the high-temperature thermal gradient (104-105°C/cm) [10].

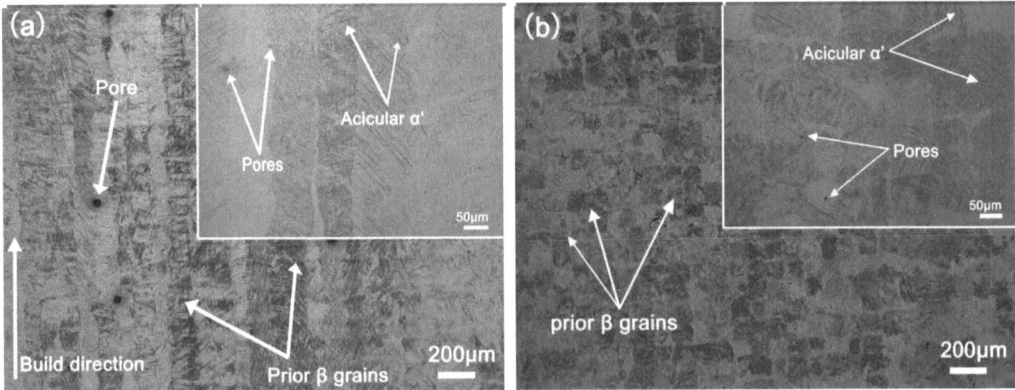

Fig. (7). Optical microstructure of the (**a**) XZ-plane and (**b**) XY-plane of SLM-produced Ti-6Al-4V sample. The insets is the enlarge images corresponding to each plane of the SLM-produced alloy [38].

CORROSION BEHAVIOR OF SLM-PRODUCED TI-6AL-4V AFTER HEAT TREATMENT

As seen from the previous Section, the SLM-produced Ti-6Al-4V alloy exhibits a high-energy metastable α' martensite with respect to α martensite compared to the conventionally processed grade 5 alloy; there also exists a slight difference in corrosion resistance. If α' martensite in Ti-6Al-4V manufactured by SLM can be transformed to α martensite by secondary processing like heat treatment, in theory, its corrosion resistance might be improved. For this purpose, the α' martensite was then transformed to α martensite by heat treatment at 500, 850 or 1000°C for 2 hours in a tube furnace, in order to confirm the α' martensite was transferred to β phase entirely. The optical microstructures (Fig. 8) of heat-treated Ti-6Al-4V produced by SLM method will help to recognize. It can be seen from this figure that the distribution of acicular α' martensite in the whole microstructure is overwhelming. Accompanied by long, columnar, previous β grains and grown along the construction direction. Specifically, the fine α' martensite can still be observed inside the primary β grains, as seen from the enlarged image in the illustration of Fig. (**8a**). Some pores can also be trapped in the microstructure. A detailed description of the microstructures of Ti-6Al-4V alloys produced by SLM can be found in references [35, 38]. An explanation of the formation of primary β grains and the existence of pore defects have been discussed previously [37]. The primary β-grain and fine α' martensite phases are still visible in the microstructure. It is worth noting that the α' martensite phase is being transformed into a plate-like α phase. The microstructure of the sample (temperature is still lower than T_β) is heat treated at 850°C (Fig. **8c**). Fine needle-like α' martensite and columnar β grains are disappeared. Instead, they are transformed to plate-like α-phase and β-phase grains. Since the β-transition temperature (T_β) of the Ti-6Al-4V alloy is about 995°C, when the heat treatment is carried out above T_β [36], a uniform 100% β phase is formed in the microstructure, and the layered α + β mixture is formed during furnace cooling. For the heat treatment temperature of 1000°C (*i.e.* more than T_β), the microstructure of the heat-treated sample is basically the same as that of T_β. The long column β phase disappears, indicating that the grain of the SLM production alloy grows up to the equiaxed β grains. On the other hand, the mixture of lamellar α + β microstructures is formed by cooling in the furnace. As shown in the enlarged view of the illustration in Fig. (**8d**), the bright region corresponds to the α phase and the dark region belongs to the β phase of the layered transition. Similar results of the heat-treated effects on the microstructure of SLM-produced Ti-6Al-4V alloys were obtained by Vilaro *et al.* [47] and Vrancken *et al.* [36].

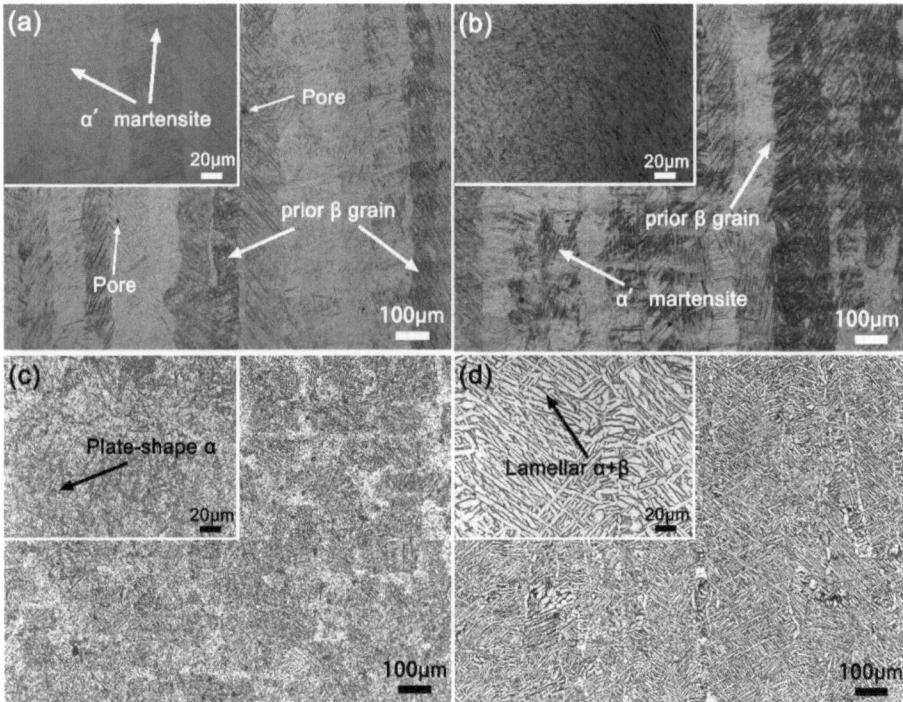

Fig. (8). Optical microstructure of the SLM-produced Ti-6Al-4V alloy: (**a**) as-received sample, and heat-treated samples at (**b**) 500 °C, (**c**) 850 °C and (**d**) 1000 °C (beyond T_β). The insets are the enlarge images for corresponding samples [43].

Fig. (**9a**) displays the X-ray Diffraction (XRD) patterns of the as-received SLM-produced sample; the main peaks from α' martensite phase are observed. It has been pointed out that, in the Ti-6Al-4V alloy produced by SLM, the β-Ti phase is hard to detect or show a few or even be absent [13, 36]. It is also suggested that the peak near $2\theta = 72°$ may be α-Ti or β-Ti phase [48 - 50]. For a sample heat treated at 500°C, the main peak remains the same as the sample to be tested

(Fig. **9b**). The (110) β-Ti peak is present in the XRD pattern when heat treated at 850°C. At the same time, the peaks near $2\theta = 35°$, 39°, 41°, 53.5°, 64°, 77° and 78° correspond to the plate-like α phase. As shown in Fig. (**9c**), the elongated plate is an α phase. More clearly, the peak of β-Ti near $2\theta = 39.5°$ in the XRD pattern indicates that the amount of β-Ti phase is produced in the microstructure and can be examined by XRD. It can be seen from Table **2** that the phase composition and volume fraction (V_f) of the SLM produced Ti-6Al-4V alloy can be analyzed from the XRD patterns by Jade 5.0 software. Methods for calculating the microstructure phase content are described elsewhere [9, 11, 51]. The volume fraction of the β phase in the microstructure of the original sample and heat treated to 500°C is very close, while the volume fraction of the β phase is significantly increased at heat treatment of 850°C and 1000°C. In addition, the peak of the α-Ti phase exhibits a strong intensity, which is related to the increase in grain size. The results of the XRD pattern are in good agreement with the results of the microstructure.

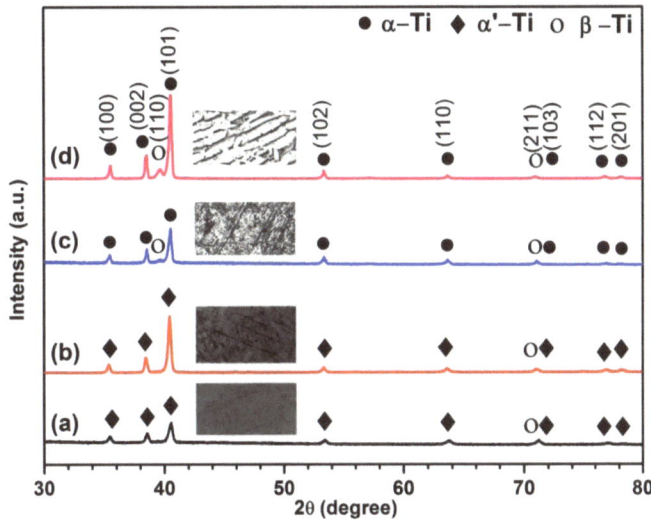

Fig. (9). XRD patterns of the SLM-produced Ti-6Al-4V alloy: (**a**) as-received sample, and heat-treated samples at (**b**) 500 °C, (**c**) 850 °C and (**d**) 1000 °C (beyond T_β) [43].

Table 2. Phase constituents and their volume fraction (V_f) of as-received and heat-treated SLM-produced Ti-6Al-4V alloy, calculated from XRD patterns [43].

Sample	Phase constituent	$V_{f,\alpha}$ or $V_{f,\alpha'}$	$V_{f,\beta}$
As-received	α' + β	95.0%	5.0%
Heat treated at 500 °C	α' + β	95.3%	4.7%
Heat treated at 850 °C	α + β	89.1%	10.9%
Heat treated at 1000 °C	α + β	87.9%	12.1%

Fig. (**10**) shows the OCP curve of the Ti-6Al-4V alloy prepared from an untreated and heat-treated samples immersed in a 3.5 wt% NaCl solution. Both OCP curves for untreated and heat-treated samples show positive movement as the immersion time prolongs, which indicates that the passivation film (TiO$_2$) is formed on the sample surface. For the samples with different heat treatment temperatures, it takes about 50 hours to obtain a relatively stable OCP value. After 50 hours of immersion, the change rate of OCP is 1.5 mV/h, which has no significant effect on continuous EIS detection. The final OCP value of the unheat-treated Ti-6Al-4V alloy produced by SLM is -79 ± 9.5 mV. The OCP of the SLM produced Ti-6Al-4V alloy samples with heat treated at 500°C and 850°C has a lower value than the untreated samples. The OCP values of the samples heat treated at 500°C and 850°C are -115 ± 10.6 mV and -155 ± 12.1 mV, respectively. For the samples subjected to heat treatment at 1000°C, OCP show a maximum negative value of -175 ± 1.7 mV.

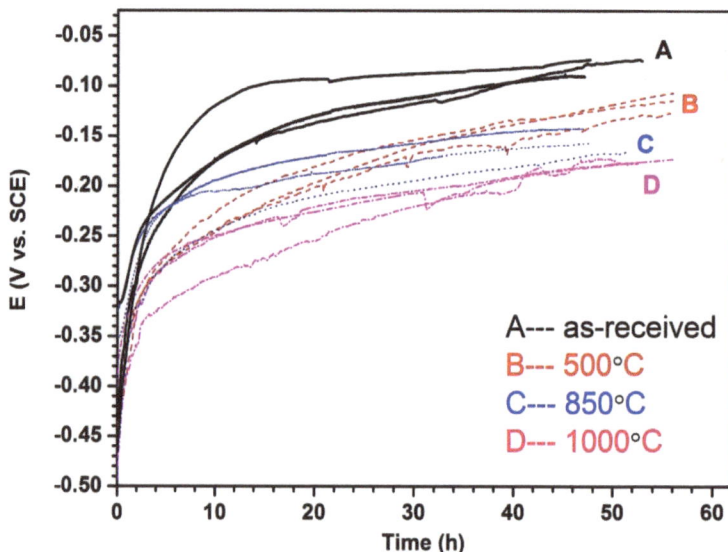

Fig. (10). Open circuit potential of as-received and heat-treated SLM-produced Ti-6Al-4V alloys in 3.5 wt.% NaCl solution [43].

Fig. (**11**) shows the potentiodynamic polarization curves for untreated and heat treated SLM Ti-6Al-4V alloys. The fitting values of the corrosion current density (i_{corr}) and the passivation current density (i_p) are based on the polarization curve of Fig. (**11**) are summarized in Table **3**. For untreated samples, the clear passivation behavior is within the range of 650 mV to 1200 mV; the i_p of the untreated alloy is about 0.9 ± 0.04 μA·cm^{-2} ($i_{p,A}$). In addition, a second passivation behavior is observed at the potential of 1600 mV. In general, the passivation current density (i_p) describes the stability of the passivation film and shows how much the current density can support the film generation period. When the sample is heat treated at 500°C, the polarization curve also shows a passivation characteristic in the potential range between 600 mV and 1250 mV; the passivation current density ($i_{p,B}$) is 1.3 ± 0.07 μA·cm^{-2}, which is slightly higher than that for the untreated sample. This indicates that the samples heat treated at 500°C exhibit poor passivation behavior compared to non-heat-treated samples. Likewise, the alloy sample heat-treated at 500 °C also has a second passivation of a potential of after 1500 mV; however, the value of the second passivation current density appears to be lower than that for untreated sample. The Ti-6Al-4V alloy produced by SLM heat treated at 850 °C shows a similar electrode behavior compared to the sample treated at 500°C; it appeared that the sample exhibits a higher passivation current density ($i_{p,C}$) of 1.5 ± 0.05 μA·cm^{-2}. For the samples that are heat treated beyond the β-transus temperature at 1000°C, the polarization curve no longer has good passivation properties throughout the anode branch.

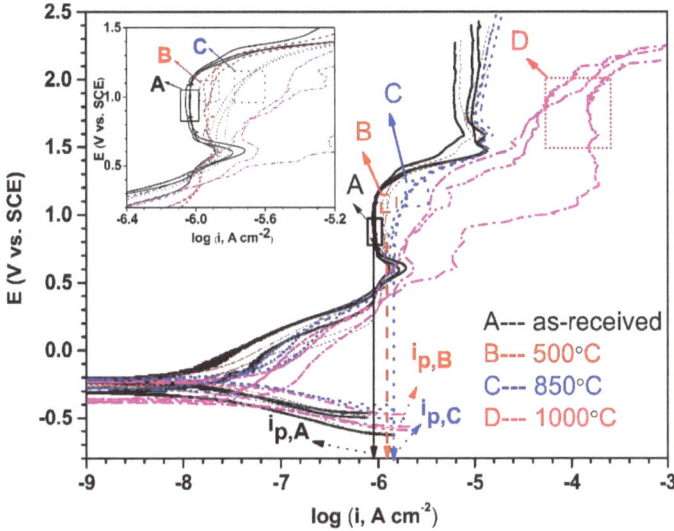

Fig. (11). Potentiodynamic polarization curves of the as-received and heat-treated SLM-produced Ti-6Al-4V alloy samples in 3.5 wt.% NaCl solution [43].

Table 3. Fitted values of corrosion current density (i_{corr}) and passive current density (i_p) of the as-received and heat-treated SLM-produced Ti-6Al-4V alloys [43].

Samples	As-received	500 °C	850 °C	1000 °C
i_{corr} (nA cm^{-2})	13.1 (±2.5)	14.6 (±1.1)	55.8 (±3.0)	76.1 (±8.3)
i_p (μA cm^{-2})	0.9 (±0.04)	1.3 (±0.07)	1.5 (±0.05)	--

In addition, to assess the corrosion rate of different samples, the corrosion rate (i_{corr}) of the heat-treated sample is calculate. Atapour *et al.* [52, 53] and Yilbas *et al.* [54] evaluated the corrosion rate of titanium alloys by i_{corr}. The fitting results of i_{corr} are listed in Table **3**. It is clear that the heat-treated sample has a higher corrosion rate than the untreated sample; the higher the heat treatment temperature, the higher the corrosion rate (i_{corr}). Therefore, the heat treatment cannot improve the corrosion resistance of Ti-6Al-4V alloy produced by SLM.

The EIS measurements were performed to further study the interface information and an electrochemical process of the as-received and heat-treated SLM-produced Ti-6Al-4V alloys in the 3.5 wt.% NaCl solution. Fig. (**12**) shows the Bode plot of the test samples, and the equivalent circuit in Fig. (**12a**) is used to fit the EIS data. For EIS data, the Nyquist plot (not shown here) exhibits only one large capacitive loop, while the Bode plot shows an extensive flattening from high frequency to low frequency. So, one time-constant equivalent circuit will be use to fit the EIS data. To represent this equivalent circuit, R_s represents the solution resistance, and R_{ct} represents the charge transfer resistance. The constant phase element (*CPE*) corresponds to the double layer capacitance. The results of the EIS fitting data for untreated and heat-treated Ti-6Al-4V alloy immersed in 3.5 wt% NaCl solution are shown in Table **4**. Fig. (**12**) also shows the fitting results of the untreated and heated R_{ct}. The R_{ct} of the untreated sample also shows a very high value, while the R_{ct} value of the heat-treated sample is significantly reduced. This indicates that the Ti-6Al-4V alloy produced by post-heat treatment exhibits a high corrosion rate. The results of the polarization curve are strongly consistent with the results for the EIS test. In addition, the fitting results of Q (*CPE*, constant phase element) and n are also shown in Table **4**. According to the n value close to 1, the electric double layer has a very high capacitance characteristic. The double-layer capacitance for un-heat-treated and heat-treated also exhibit a very close value.

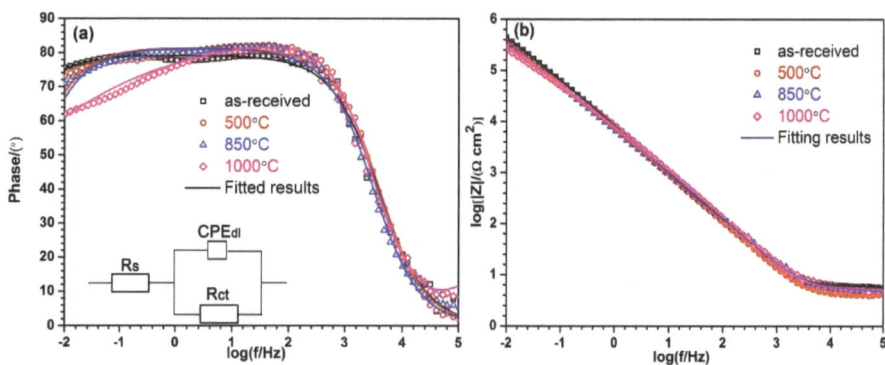

Fig. (12). Bode plots (**a**) and (**b**) of EIS for the as-received and heat-treated SLM-produced Ti-6Al-4V alloys in 3.5 wt.% NaCl solution: The inserted equivalent circuit in (**a**) is used to fit the impedance spectra [43].

Table 4. The fitted results of EIS measurements for the as-received and heat treated SLM-produced Ti-6Al-4V alloys in 3.5 wt.% NaCl solution [43].

Sample	R_s ($\Omega \cdot cm^2$)	CPE_{dl}, Y_0 10^{-4} ($S \cdot Sec^n \cdot cm^{-2}$)	n_1	R_{ct} ($M\Omega \cdot cm^2$)
As-received	4.7 (±0.38)	1.48 (±0.13)	0.8857	4.7 (±0.74)
500°C	3.8 (±0.69)	1.38 (±0.06)	0.9086	2.4 (±0.25)
850°C	5.3 (±0.70)	1.26 (±0.10)	0.8962	1.8 (±0.07)
1000°C	4.9 (±0.17)	1.31 (±0.04)	0.8925	0.6 (±0.17)

CORROSION BEHAVIOR OF SLM-PRODUCED TITANINUM COMPOSITE

SLM-produced Ti-TiB composites can be considered one of the most suitable candidates for high strength and anti-wear biomedical applications [55]. Also, the porous structure of Ti-TiB composites is conductive to bone cell growth and also reduces the risk of rejection of the substitutes in patients [56, 57].

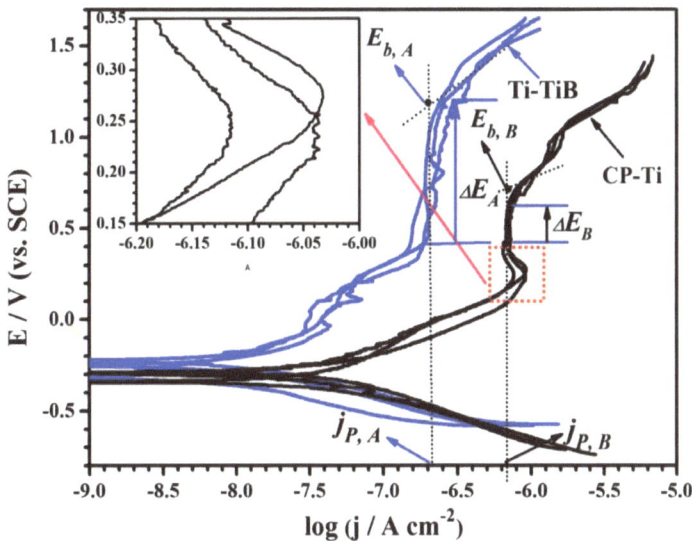

Fig. (13). The polarization curves of CP-Ti and Ti-TiB samples produced by SLM immersed in aerated Hank's solution at body temperature [60].

Fig. (13) shows the typical potential polarization curves for CP-Ti and Ti-TiB samples produced by SLM technique in Hank solution at body temperature. Table 5 is the fitted parameters obtained from the polarization curve, such as passive region range (ΔE), passivation current density (j_p) and breakdown potential (E_b). A typical passivation zone is observed in the curves of CP-Ti and

Ti-TiB samples. In particular, the Ti-TiB sample has a relatively large passivation zone (0.40 V to 1.25 V) compared to the CP-Ti sample (0.40 V to 0.70 V). With regard to the passivation current density (j_p), the average Ti-TiB sample is about 0.22 µA cm^{-2}, which is much lower than that of the CP-Ti sample (0.75 µA cm^{-2}). As a general rule, j_p represents the passivation current of the film formed on the metal surface, and the smaller j_p value indicates that the sample is susceptible to passivation [58]. In addition, the breakdown potential (E_b) is used to assess the sensitivity of the passivation film to halogen ion damage. For the breakdown potential (E_b), the larger value of E_b represents a more stable passivation film formed on the metal surface [59]. Thus, the Ti-TiB sample has a larger E_b value compared to the CP-Ti sample, indicating that the Ti-TiB sample produced by the SLM technique has better corrosion resistance than the CP-Ti sample.

Table 5. Corrosion parameters obtained from polarization curves of CP-Ti and Ti-TiB samples produced by SLM immersed in aerated Hank's solution at body temperature [60].

Sample	j_p /µA cm^{-2}	E_b / V (*vs.* SCE)	ΔE / V (*vs.* SCE)
CP-Ti	0.75	0.70	0.30
Ti-TiB	0.22	1.23	0.79

The Fig. (**13**) inset shows that the CP-Ti sample exhibits an active-passive behavior during the anodic polarization when the potential is in the range of 0.15V to 0.35V, indicating an incomplete passive behavior at the OCP [61]. After that, the CP-Ti samples exhibited stable passive behavior and increased potential. In this segmented anodic polarization process, due to the presence of Cl$^-$ [62]. The CP-Ti sample may produce a large amount of Ti^{4+} in the first step. The released Ti^{4+} is then transported to the electrolyte and reacted with a Hank solution (*e.g.*, Cl$^-$) [63]. Finally, when the concentration of the [TiCl$_6$]$^{2-}$ complex reaches a critical value, the previously formed [TiCl$_6$]$^{2-}$ complex is hydrolyzed and then a TiO$_2$ passivation film is formed on the surface of the metal-electrolyte, which prevents The Ti matrix and the corrosive medium, thereby reducing the tendency to corrosion (Eq. 1) [63, 64]:

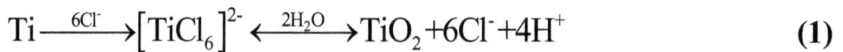

$$\text{Ti} \xrightarrow{\text{6Cl}^-} \left[\text{TiCl}_6\right]^{2-} \xleftrightarrow{\text{2H}_2\text{O}} \text{TiO}_2 + 6\text{Cl}^- + 4\text{H}^+ \qquad (1)$$

However, the Ti-TiB samples produced by SLM methods do not show similar behavior in the anodic polarization, but directly into the passivation zone, which confirmed the above-mentioned easy passivation. As seen from the SEM images after potentiodynamic polarization experiments, for SLM-produced CP-Ti samples (Fig. **14**), many pits and large amounts of corrosion products are observed throughout and the size of pits range from nanometer to micron. In contrast, as shown in Fig. (**14b**), SLM-produced Ti-TiB samples are covered with a passive film on the surface that effectively inhibits the Cl⁻ ion attack of the titanium matrix: TiO_2 passivation film (grey).

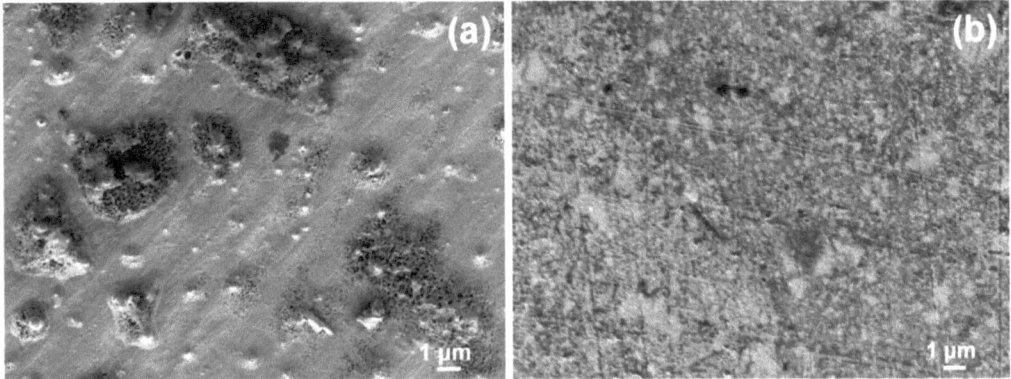

Fig. (14). SEM images of (**a**) SLM-produced CP-Ti samples and (**b**) SLM-produced Ti-TiB composite samples after polarization tests in aerated Hank's solution at body temperature [60].

Fig. (**15a-c**) represents the Nyquist and Bode plots for the SLM-produced CP-Ti and Ti-TiB samples in aerated Hank's solution at body temperature. From the Nyquist plot (Fig. **15a**), it is clear that for this two samples, only one large single capacitor arc is visible. However, the Bode plots (Fig. **15c**) show a platform from intermediate to low frequencies. Thus, in Fig. (**15d**), an equivalent circuit consisting of two-time constants is used to simulate the corrosion behavior in order to better fit the EIS dates of CP-Ti and Ti-TiB samples. In this equivalent circuit, R_s corresponds to the solution resistance; R_f describes the resistance of the passive film formed on the metal surface; R_{ct} reflects the charge transfer resistance; and CPE_1 and CPE_2 are constant phase elements, representing the deviation from ideal capacitive behavior of the passive film and double-layer, respectively, which can be attributed to roughness or defects of the electrode surface [64, 65].

Table **6** lists the fitting results of the EIS measurements. Chi-square values (χ^2) between 2.4×10^{-3} and 9.2×10^{-3} indicates that the good quality of obtained EIS data. It can be seen from the fitting results that the average of the Ti-TiB samples (2.05 MΩ cm²) produced by the SLM is significantly larger than the average of the CP-Ti samples (0.31 MΩ cm²). In general, a higher R_f value means that the corrosion resistance of the sample is better [66]. The higher average R_{ct} of the SLM produced Ti-TiB samples (1.44 MΩ cm²) compared to the SLM-produced CP-Ti samples (0.54 MΩ cm²) also supports this conclusion.

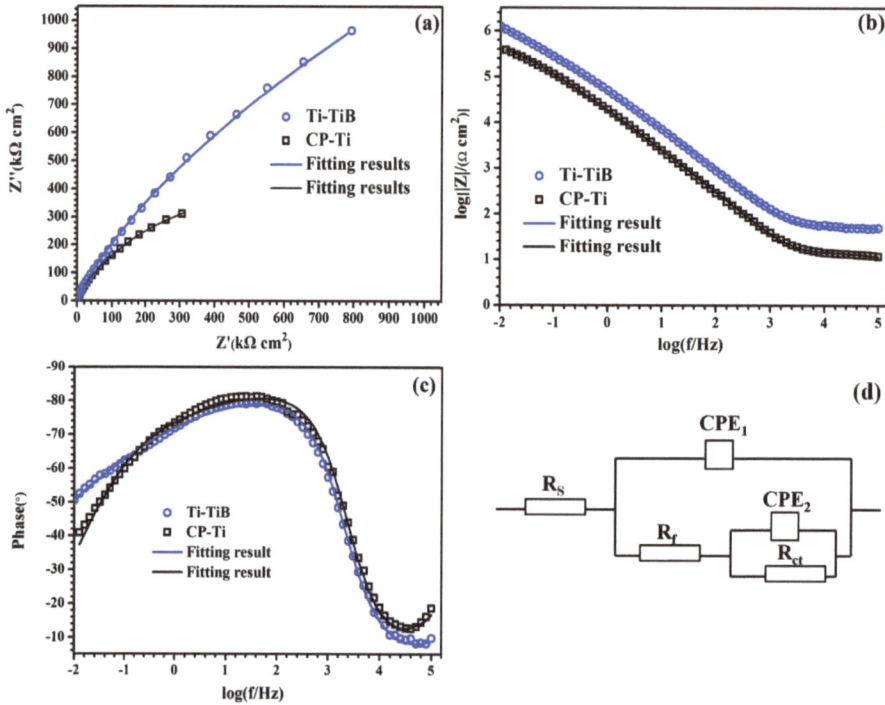

Fig. (15). Electrochemical impedance spectroscopy for CP-Ti and Ti-TiB samples produced by SLM immersed in aerated Hank's solution at body temperature: (**a**) Nyquist plots (**b**) and (**c**) Bode plots and (**d**) equivalent circuit model [60].

Table 6. Fitting results of EIS for the CP-Ti and Ti-TiB samples produced by SLM immersed in aerated Hank's solution at body temperature [60].

Sample	R_f (MΩ cm^2)	CPE$_1 \times 10^{-6}$ (Ω^{-1} cm^{-2} S^{-n})	n_1	R_{ct} (MΩ cm^2)	CPE$_{dl} \times 10^{-6}$ (Ω^{-1} cm^{-2} S^{-n})	n_2	$\chi^2 \times 10^{-3}$
CP-Ti	0.31	12.10	0.86	0.54	17.03	0.82	7.0
Ti-TiB	2.05	3.34	0.90	1.44	5.14	0.86	6.3

Fig. (**16**) shows SEM images of CP-Ti and Ti-TiB samples after 7 weeks of immersion in Hank solution at body temperature. Apparently, both samples have the same corrosion products. Fig. (**16b**) shows the whole process of these corrosion products from the spherical product to the collected mass. Some previous studies have explained that these titanium corrosion products can cause Cl$^-$ to destroy the TiO$_2$ passivation film and even induce pitting on the titanium substrate [67, 68]. In Fig. (**16a and c**), CP-Ti samples have more corrosion products compared to Ti-TiB samples, indicating that CP-Ti produced by SLM is more susceptible to corrosion than Ti-TiB produced by SLM.

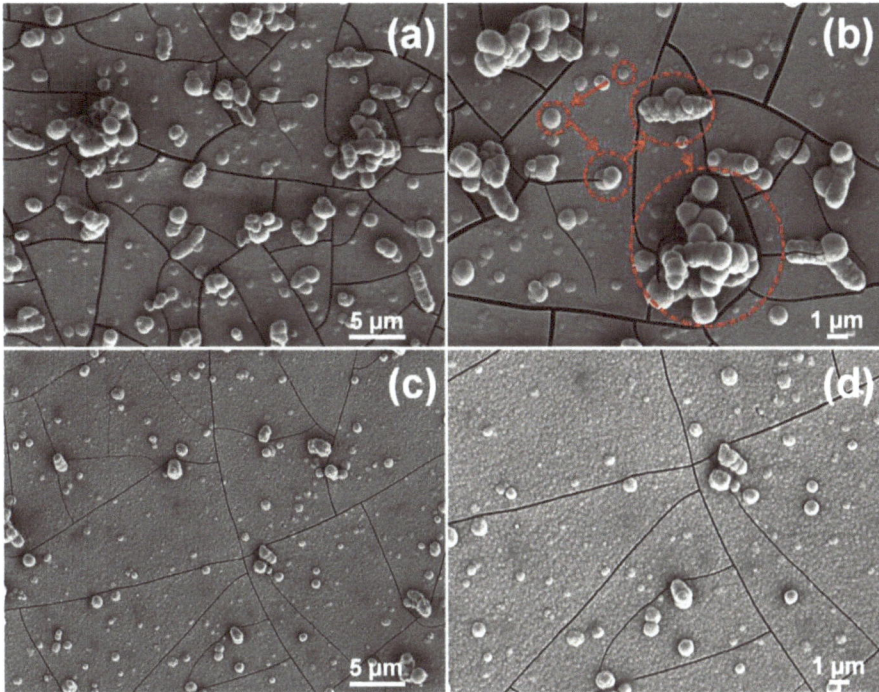

Fig. (16). SEM images of samples after immersing in Hank's solution for 7 weeks at body temperature for (**a**) and (**b**) SLM-produced CP-Ti samples, (**c**) and (**d**) SLM-produced Ti-TiB composite samples [60].

CONCLUDING REMARKS

This chapter has reviewed the corrosion behavior of both Ti-6Al-4V and Ti-TiB composite manufactured by selective laser melting (SLM). Electrochemical results show that the SLM-produced sample possesses slightly inferior corrosion resistance than the Grade 5 sample. Microstructure studies suggest that the SLM-produced sample is composed of dominant acicular α' martensite and some primary β grains, unlike the typical α+β microstructure in Grade 5 sample. The unfavorable corrosion resistance of the SLM-produced sample is related to the considerably large amount of acicular α' and less β-Ti phase in the microstructure compared to the Grade 5 sample. However, the transformation of α' martensite to α martensite by heat treatment in SLM-produced Ti-6Al-4V alloy could not further improve its corrosion resistance. SLM-produced Ti-TiB composite samples possess better corrosion resistance than SLM-produced CP-Ti samples in Hank's solution. Due to these tiny TiB and TiB_2 particles acting as the micro-cathode uniformly distributing in titanium matrix, anodic dissolution of titanium matrix in the corrosion process is prominently facilitated in early stages, followed by rapid passivation on the surface.

CONFLICT OF INTEREST

The author (editor) declares no conflict of interest, financial or otherwise.

ACKNOWLEDGEMENTS

This research was supported by the Australian Research Council's Discovery Projects (DP110101653) and by the Project of Shanghai Science and Technology Commission (14DZ2261000). The authors are grateful to N.W. Dai, Y. Chen, X.H. Gu, P. Duan for their collaboration.

REFERENCES

[1]   Handzlik P, Fitzner K. Corrosion resistance of Ti and Ti-Pd alloy in phosphate buffered saline solutions with and without H_2O_2 addition. J Nonferr Metal Soc 2013; 23: 866-75.
       [http://dx.doi.org/10.1016/S1003-6326(13)62541-8]

[2]   Marino CE. Oliveira EMd, Rocha-Filho RC, Biaggio SR. On the stability of thin-anodic-oxide films of titanium in acid phosphoric media. Corros Sci 2001; 43: 1465-76.
       [http://dx.doi.org/10.1016/S0010-938X(00)00162-1]

[3]   Chen JR, Tsai WT. *In situ* corrosion monitoring of Ti-6Al-4V alloy in H_2SO_4/HCl mixed solution using electrochemical AFM. Electrochim Acta 2011; 56: 1746-51.
       [http://dx.doi.org/10.1016/j.electacta.2010.10.024]

[4]   De Assis SL, Wolynec S, Costa I. Corrosion characterization of titanium alloys by electrochemical techniques. Electrochim Acta 2006; 51: 1815-9.
       [http://dx.doi.org/10.1016/j.electacta.2005.02.121]

[5]   Long M, Rack HJ. Titanium alloys in total joint replacement--a materials science perspective. Biomaterials 1998; 19(18): 1621-39.
       [http://dx.doi.org/10.1016/S0142-9612(97)00146-4] [PMID: 9839998]

[6]   He G, Eckert J, Dai QL, *et al.* Nanostructured Ti-based multi-component alloys with potential for biomedical applications. Biomaterials 2003; 24(28): 5115-20.
       [http://dx.doi.org/10.1016/S0142-9612(03)00440-X] [PMID: 14568427]

[7]   Brewer WD, Bird RK, Wallace TA. Titanium alloys and processing for high speed aircraft. Mater Sci Eng A 1998; 243: 299-304.
       [http://dx.doi.org/10.1016/S0921-5093(97)00818-6]

[8]   Cho G-B, Kim K-W, Ahn H-J, Cho K-K, Nam T-H. Applications of Ti-Ni alloys for secondary batteries. J Alloys Compd 2008; 449: 317-21.
       [http://dx.doi.org/10.1016/j.jallcom.2006.01.129]

[9]   Haghighi SE, Lu HB, Jian GY, Cao GH, Habibi D, Zhang LC. Effect of α'' martensite on the microstructure and mechanical properties of beta-type Ti-Fe-Ta alloys. Mater Des 2015; 76: 47-54.
       [http://dx.doi.org/10.1016/j.matdes.2015.03.028]

[10]   Zhang LC, Attar H. Selective laser melting of titanium alloys and titanium matrix composites for biomedical applications: a review. Adv Eng Mater 2016; 18: 463-75.
       [http://dx.doi.org/10.1002/adem.201500419]

[11]   Ehtemam-Haghighi S, Liu Y, Cao G, Zhang L-C. Influence of Nb on the $\beta \rightarrow \alpha''$ martensitic phase transformation and properties of the newly designed Ti-Fe-Nb alloys. Mater Sci Eng C 2016; 60: 503-10.
       [http://dx.doi.org/10.1016/j.msec.2015.11.072] [PMID: 26706557]

[12] Pohrelyuk I, Fedirko V, Tkachuk O, Proskurnyak R. Corrosion resistance of Ti-6Al-4V alloy with nitride coatings in Ringer's solution. Corros Sci 2013; 66: 392-8.
[http://dx.doi.org/10.1016/j.corsci.2012.10.005]

[13] Amaya-Vazquez M, Sánchez-Amaya J, Boukha Z, Botana F. Microstructure, microhardness and corrosion resistance of remelted TiG2 and Ti6Al4V by a high power diode laser. Corros Sci 2012; 56: 36-48.
[http://dx.doi.org/10.1016/j.corsci.2011.11.006]

[14] Piazza S, Biundo GL, Romano MC, Sunseri C, Di Quarto F. *In situ* characterization of passive films on Al-Ti alloy by photocurrent and impedance spectroscopy. Corros Sci 1998; 40: 1087-108.
[http://dx.doi.org/10.1016/S0010-938X(98)00009-2]

[15] Alves VA, Reis RQ, Santos IC, *et al. In situ* impedance spectroscopy study of the electrochemical corrosion of Ti and Ti-6Al-4V in simulated body fluid at 25°c and 37°C. Corros Sci 2009; 51: 2473-82.
[http://dx.doi.org/10.1016/j.corsci.2009.06.035]

[16] González JE, Mirza-Rosca JC. Study of the corrosion behavior of titanium and some of its alloys for biomedical and dental implant applications. J Electroanal Chem 1999; 471: 109-15.
[http://dx.doi.org/10.1016/S0022-0728(99)00260-0]

[17] Marino CE, Mascaro LH. EIS characterization of a Ti-dental implant in artificial saliva media: dissolution process of the oxide barrier. J Electroanal Chem 2004; 568: 115-20.
[http://dx.doi.org/10.1016/j.jelechem.2004.01.011]

[18] Ibriş N, Rosca JC. EIS study of Ti and its alloys in biological media. J Electroanal Chem 2002; 526: 53-62.
[http://dx.doi.org/10.1016/S0022-0728(02)00814-8]

[19] Razavi RS, Salehi M, Ramazani M, Man HC. Corrosion Behaviour of laser gas nitrided Ti-6Al-4V in hcl solution. Corros Sci 2009; 51: 2324-9.
[http://dx.doi.org/10.1016/j.corsci.2009.06.016]

[20] Galvanetto E, Galliano F, Fossati A, Borgioli F. Corrosion resistance properties of plasma nitrided Ti-6Al-4V alloy in hydrochloric acid solutions. Corros Sci 2002; 44: 1593-606.
[http://dx.doi.org/10.1016/S0010-938X(01)00157-3]

[21] Heakal FE, Ghoneim A, Mogoda A, Awad K. Electrochemical behaviour of Ti-6Al-4V alloy and Ti in azide and halide solutions. Corros Sci 2011; 53: 2728-37.
[http://dx.doi.org/10.1016/j.corsci.2011.05.003]

[22] Geetha M, Mudali UK, Gogia A, Asokamani R, Raj B. Influence of microstructure and alloying elements on corrosion behavior of Ti-13Nb-13Zr alloy. Corros Sci 2004; 46: 877-92.
[http://dx.doi.org/10.1016/S0010-938X(03)00186-0]

[23] Zhang LC, Klemm D, Eckert J, Hao YL, Sercombe TB. Manufacture by selective laser melting and mechanical behavior of a biomedical Ti-24Nb-4Zr-8Sn alloy. Scr Mater 2011; 65: 21-4.
[http://dx.doi.org/10.1016/j.scriptamat.2011.03.024]

[24] Wang Q, Liu Z, Wang B, Hassan Mohsan AU. Stress-induced orientation relationship variation for phase transformation of α-Ti to β-Ti during high speed machining Ti-6Al-4V. Mater Sci Eng A 2017; 690: 32-6.
[http://dx.doi.org/10.1016/j.msea.2017.02.098]

[25] Odenberger E-L, Pederson R, Oldenburg M. Thermo-mechanical material response and hot sheet metal forming of Ti-6242. Mater Sci Eng A 2008; 489: 158-68.
[http://dx.doi.org/10.1016/j.msea.2007.12.047]

[26] Liu J, Chang L, Liu H, Li Y, Yang H, Ruan J. Microstructure, mechanical behavior and biocompatibility of powder metallurgy Nb-Ti-Ta alloys as biomedical material. Mater Sci Eng C 2017; 71: 512-9.

[http://dx.doi.org/10.1016/j.msec.2016.10.043] [PMID: 27987739]

[27]   Attar H, Calin M, Zhang LC, Scudino S, Eckert J. Manufacture by selective laser melting and mechanical behavior of commercially pure titanium. Mater Sci Eng A 2014; 593: 170-7.
[http://dx.doi.org/10.1016/j.msea.2013.11.038]

[28]   Liu Y, Li S, Hou W, *et al.* Electron beam melted beta-type Ti-24Nb-4Zr-8Sn porous structures with high strength-to-modulus ratio. J Mater Sci Technol 2016; 32: 505-8.
[http://dx.doi.org/10.1016/j.jmst.2016.03.020]

[29]   Zhang L, Sercombe T. Selective laser melting of low-modulus biomedical Ti-24Nb-4Zr-8Sn alloy: effect of laser point distance. Key Eng Mater 2012; 520: 226-33.
[http://dx.doi.org/10.4028/www.scientific.net/KEM.520.226]

[30]   Liu YJ, Li SJ, Wang HL, *et al.* Microstructure, defects and mechanical behavior of beta-type titanium porous structures manufactured by electron beam melting and selective laser melting. Acta Mater 2016; 113: 56-67.
[http://dx.doi.org/10.1016/j.actamat.2016.04.029]

[31]   Attar H, Prashanth K, Chaubey A, *et al.* Comparison of wear properties of commercially pure titanium prepared by selective laser melting and casting processes. Mater Lett 2015; 142: 38-41.
[http://dx.doi.org/10.1016/j.matlet.2014.11.156]

[32]   Attar H, Löber L, Funk A, *et al.* Mechanical behavior of porous commercially pure Ti and Ti-TiB composite materials manufactured by selective laser melting. Mater Sci Eng A 2015; 625: 350-6.
[http://dx.doi.org/10.1016/j.msea.2014.12.036]

[33]   Attar H, Bönisch M, Calin M, Zhang L-C, Scudino S, Eckert J. Selective laser melting of *in situ* titanium-titanium boride composites: processing, microstructure and mechanical properties. Acta Mater 2014; 76: 13-22.
[http://dx.doi.org/10.1016/j.actamat.2014.05.022]

[34]   Liu YJ, Li XP, Zhang LC, Sercombe TB. processing and properties of topologically optimised biomedical Ti-24Nb-4Zr-8Sn scaffolds manufactured by selective laser melting. Mater Sci Eng A 2015; 642: 268-78.
[http://dx.doi.org/10.1016/j.msea.2015.06.088]

[35]   Dai N, Zhang LC, Zhang J, Chen Q, Wu M. Corrosion behavior of selective laser melted Ti-6Al-4 V alloy in nacl solution. Corros Sci 2016; 102: 484-9.
[http://dx.doi.org/10.1016/j.corsci.2015.10.041]

[36]   Vrancken B, Thijs L, Kruth J-P, Van Humbeeck J. Heat treatment of Ti6Al4V produced by Selective Laser Melting: Microstructure and mechanical properties. J Alloys Compd 2012; 541: 177-85.
[http://dx.doi.org/10.1016/j.jallcom.2012.07.022]

[37]   Thijs L, Verhaeghe F, Craeghs T, Van Humbeeck J, Kruth J-P. A study of the microstructural evolution during selective laser melting of Ti-6Al-4V. Acta Mater 2010; 58: 3303-12.
[http://dx.doi.org/10.1016/j.actamat.2010.02.004]

[38]   Dai N, Zhang LC, Zhang J, *et al.* Distinction in corrosion resistance of selective laser melted Ti-6Al-4V alloy on different planes. Corros Sci 2016; 111: 703-10.
[http://dx.doi.org/10.1016/j.corsci.2016.06.009]

[39]   Bai Y, Gai X, Li S, *et al.* Improved corrosion behaviour of electron beam melted Ti-6Al-4V alloy in phosphate buffered saline. Corros Sci 2017; 123: 289-96.
[http://dx.doi.org/10.1016/j.corsci.2017.05.003]

[40]   Wu S, Lu Y, Gan Y, *et al.* Microstructural evolution and microhardness of a selective-laser-melted Ti-6Al-4V alloy after post heat treatments. J Alloys Compd 2016; 672: 643-52.
[http://dx.doi.org/10.1016/j.jallcom.2016.02.183]

[41]   Huang Q, Liu X, Yang X, Zhang R, Shen Z, Feng Q. Specific heat treatment of selective laser melted Ti-6Al-4V for biomedical applications. Front Mater Sci China 2015; 9: 373-81.

[http://dx.doi.org/10.1007/s11706-015-0315-7]

[42] Wauthle R, Vrancken B, Beynaerts B, *et al.* Effects of build orientation and heat treatment on the microstructure and mechanical properties of selective laser melted Ti6Al4V lattice structures. Add Manuf 2015; 5: 77-84.

[43] Dai N, Zhang J, Chen Y, Zhang L-C. Heat Treatment Degrading the Corrosion Resistance of Selective Laser Melted Ti-6Al-4V Alloy. J Electrochem Soc 2017; 164: C428-34.
       [http://dx.doi.org/10.1149/2.1481707jes]

[44] Karimzadeh F, Heidarbeigy M, Saatchi A. Effect of heat treatment on corrosion behavior of Ti-6Al-4V alloy weldments. J Mater Process Technol 2008; 206: 388-94.
       [http://dx.doi.org/10.1016/j.jmatprotec.2007.12.065]

[45] Yadroitsev I, Krakhmalev P, Yadroitsava I. Selective laser melting of Ti6Al4V alloy for biomedical applications: Temperature monitoring and microstructural evolution. J Alloys Compd 2014; 583: 404-9.
       [http://dx.doi.org/10.1016/j.jallcom.2013.08.183]

[46] Sallica-Leva E, Jardini AL, Fogagnolo JB. Microstructure and mechanical behavior of porous Ti-6Al-4V parts obtained by selective laser melting. J Mech Behav Biomed Mater 2013; 26: 98-108.
       [http://dx.doi.org/10.1016/j.jmbbm.2013.05.011] [PMID: 23773976]

[47] Vilaro T, Colin C, Bartout J-D. As-fabricated and heat-treated microstructures of the Ti-6Al-4V alloy processed by selective laser melting. Metall Mater Trans, A Phys Metall Mater Sci 2011; 42: 3190-9.
       [http://dx.doi.org/10.1007/s11661-011-0731-y]

[48] Sercombe T, Jones N, Day R, Kop A. Heat treatment of Ti-6Al-7Nb components produced by selective laser melting. Rapid Prototyping J 2008; 14: 300-4.
       [http://dx.doi.org/10.1108/13552540810907974]

[49] Facchini L, Magalini E, Robotti P, Molinari A, Höges S, Wissenbach K. Ductility of a Ti-6Al-4V alloy produced by selective laser melting of prealloyed powders. Rapid Prototyping J 2010; 16: 450-9.
       [http://dx.doi.org/10.1108/13552541011083371]

[50] Jovanović M, Tadić S, Zec S, Mišković Z, Bobić I. The effect of annealing temperatures and cooling rates on microstructure and mechanical properties of investment cast Ti-6Al-4V alloy. Mater Des 2006; 27: 192-9.
       [http://dx.doi.org/10.1016/j.matdes.2004.10.017]

[51] Ehtemam-Haghighi S, Liu Y, Cao G, Zhang L-C. Phase transition, microstructural evolution and mechanical properties of Ti-Nb-Fe alloys induced by Fe addition. Mater Des 2016; 97: 279-86.
       [http://dx.doi.org/10.1016/j.matdes.2016.02.094]

[52] Atapour M, Pilchak A, Shamanian M, Fathi M. Corrosion behavior of Ti-8Al-1Mo-1V alloy compared to Ti-6Al-4V. Mater Des 2011; 32: 1692-6.
       [http://dx.doi.org/10.1016/j.matdes.2010.09.009]

[53] Atapour M, Pilchak A, Frankel G, Williams J. Corrosion behavior of β titanium alloys for biomedical applications. Mater Sci Eng C 2011; 31: 885-91.
       [http://dx.doi.org/10.1016/j.msec.2011.02.005]

[54] Yilbas B, Sahin A, Ahmad Z, Aleem BA. A study of the corrosion properties of TiN coated and nitrided Ti-6Al-4V. Corros Sci 1995; 37: 1627-36.
       [http://dx.doi.org/10.1016/0010-938X(95)00065-R]

[55] Li L, Qiu F, Wang Y, *et al.* Crystalline TiB$_2$: an efficient catalyst for synthesis and hydrogen desorption/absorption performances of NaAlH4 system. J Mater Chem 2012; 22: 3127-32.
       [http://dx.doi.org/10.1039/c1jm14936a]

[56] Zhang L-C, Attar H. Selective Laser Melting of Titanium Alloys and Titanium Matrix Composites for Biomedical Applications: A Review. Adv Eng Mater 2016; 18: 463-75.
       [http://dx.doi.org/10.1002/adem.201500419]

[57]   Nelson MR, Roy K. Bone-marrow mimicking biomaterial niches for studying hematopoietic stem and progenitor cells. J Mater Chem B Mater Biol Med 2016; 4: 3490-503.
[http://dx.doi.org/10.1039/C5TB02644J]

[58]   Dai N, Zhang L, Zhang J, *et al.* Distinction in corrosion resistance of selective laser melted Ti-6Al-4V alloy on different planes. Corros Sci 2016; 111: 703-10.
[http://dx.doi.org/10.1016/j.corsci.2016.06.009]

[59]   Dai N, Zhang L-C, Zhang J, Chen Q, Wu M. Corrosion behavior of selective laser melted Ti-6Al-4V alloy in NaCl solution. Corros Sci 2016; 102: 484-9.
[http://dx.doi.org/10.1016/j.corsci.2015.10.041]

[60]   Chen Y, Zhang J, Dai N, Qin P, Attar H, Zhang L-C. Corrosion Behaviour of Selective Laser Melted Ti-TiB Biocomposite in Simulated Body Fluid. Electrochim Acta 2017; 232: 89-97.
[http://dx.doi.org/10.1016/j.electacta.2017.02.112]

[61]   Atapour M, Pilchak A, Frankel GS, Williams JC. Corrosion Behaviour Of Investment Cast And Friction Stir Processed Ti-6Al-4V. Corros Sci 2010; 52: 3062-9.
[http://dx.doi.org/10.1016/j.corsci.2010.05.026]

[62]   J.E.G. Gonzalez JCM-R. Study of the corrosion behavior of titanium and some of its alloys. Corros Sci 1999; 471: 109-15.

[63]   Saud SN, Hosseinian S R, Bakhsheshi-Rad HR, *et al.* Corrosion and bioactivity performance of graphene oxide coating on TiNb shape memory alloys in simulated body fluid. Mater Sci Eng C 2016; 68: 687-94.
[http://dx.doi.org/10.1016/j.msec.2016.06.048] [PMID: 27524069]

[64]   Popa MV, Demetrescu I, Vasilescu E, *et al.* Corrosion susceptibility of implant materials Ti-5Al-4V and Ti-6Al-4Fe in artificial extra-cellular fluids. Electrochim Acta 2004; 49: 2113-21.
[http://dx.doi.org/10.1016/j.electacta.2003.12.036]

[65]   El Taib Heakal F, Ghoneim AA, Mogoda AS, Awad K. Electrochemical behaviour of Ti-6Al-4V alloy and Ti in azide and halide solutions. Corros Sci 2011; 53: 2728-37.
[http://dx.doi.org/10.1016/j.corsci.2011.05.003]

[66]   Zaveri N, Mahapatra M, Deceuster A, Peng Y, Li L, Zhou A. Corrosion resistance of pulsed laser-treated Ti-6Al-4V implant in simulated biofluids. Electrochim Acta 2008; 53: 5022-32.
[http://dx.doi.org/10.1016/j.electacta.2008.01.086]

[67]   Rahna NB, Kalarivalappil V, Nageri M, *et al.* Stability studies of PbS sensitised TiO_2 nanotube arrays for visible light photocatalytic applications by X-ray photoelectron spectroscopy (XPS). Mater Sci Semicond Process 2016; 42: 303-10.
[http://dx.doi.org/10.1016/j.mssp.2015.10.025]

[68]   Andreatta F, Lohrengel MM, Terryn H, de Wit JH. Electrochemical characterisation of aluminium AA7075-T6 and solution heat treated AA7075 using a micro-capillary cell. Electrochim Acta 2003; 48: 3239-47.
[http://dx.doi.org/10.1016/S0013-4686(03)00379-7]

SUBJECT INDEX

A

Acicular α-phase 153
Acicular martensite 34, 35, 153
Acicular martensite phase 36
Acicular martensitic 79
Adding alloying elements 20
Aerated Hank's solution 189, 190, 191, 192
Alkaline phosphatase activity 8, 149, 157
Alloy components 94
Alloy composition 57, 67
Alloy design 25, 51
Alloying elements 6, 18, 21, 22, 48
  non-toxic 18, 48
Alloy matrix 31
Alloy microstructure 66
Alloy overmatches alloys, developed 22
Alloy porous samples 96
Alloy preparation 25
Alloy projects 19, 61
  launched commercial TiNinol 19
Alloys 3, 5, 20, 21, 22, 33, 51, 52, 57, 58, 61,
    62, 63, 64, 66, 79, 154, 186
  based 3, 52, 61, 63, 64, 66, 154
  binary 64
  bio-titanium 22
  β-phase 61, 62
  βtitanium 22
  β-titanium 20, 21, 22
  cast 79
  commercial 52
  cross-rolled 33
  deformed 33
  designed 51
  direct-rolled 33
  dual-phase 61
  magnesium 22
  metastable 5, 57
  unstable 58
  untreated 186
AM-fabricated sample 96

AM-produced titanium alloys 95
Anhydrous ethanol 135
Anisotropy 32, 40, 42
Annealed specimens 101, 102
Anodic polarization 190
Applications 2, 51, 52, 67, 73, 74, 95
  load-bearing 95
  orthopaedic 51, 52, 67
  orthopaedic surgery 52
  popular 2
  potential 73, 74
  relatively limited 74
Austenite 10, 115, 116, 120, 123
  phase B2 115
  transformation 10, 116
Average volume fraction 33, 34

B

Balling effects 78, 98
Beta titanium alloys 18, 48
Beta type Ti-Fe based alloys 51
Biomaterials, advanced implantable 1
Biomaterials alloys 73
Biomechanical properties, excellent 6
Biomedical 4, 11, 13, 18, 19, 21, 48, 114, 130,
    149, 150, 180
  field 4, 18, 19, 21, 48, 114, 149
  materials 4, 13, 18, 48, 130, 150
  Titanium Alloys by SLM 180
  titanium alloys development 11, 13
Biomedical applications 2, 3, 4, 9, 13, 51, 52,
    83, 149, 189
  anti-wear 189
Bone 8, 12, 24, 85, 130, 131, 150
  shaft healing 12
  tissue 8, 24, 85, 130, 131, 150
Brazing microstructure 129
Brittle ternary intermetallic phase 124
Bulk metallic glass (BMG) 65

www.ingramcontent.com/pod-product-compliance
Lightning Source LLC
Chambersburg PA
CBHW050840220326
41598CB00006B/417